IEE CONTROL ENGINEERING SERIES 41

Series Editors: Professor P. J. Antsaklis
Professor D. P. Atherton
Professor K. Warwick

Computer control of real-time processes

Other volumes in this series:

Computer control of real-time processes

Edited by
S. Bennett
and G. S. Virk

Peter Peregrinus Ltd. on behalf of the Institution of Electrical Engineers

Published by: Peter Peregrinus Ltd., London, United Kingdom

© 1990: Peter Peregrinus Ltd.

British Library Cataloguing in Publication Data
Computer control of real-time processes.
   1. Control systems. Applications of computer systems
   I. Bennett, Stuart   II. Virk, G. S.
   629.8312

ISBN 0 86341 220 3

Printed in England by Short Run Press Ltd., Exeter

Contents

List of contributors

Dr S Bennett
Dr G S Virk

Department of Control Engineering
University of Sheffield

Dr G C Barney

Control Systems Centre
UMIST, Manchester

Professor P J Gawthrop

Department of Mechanical Engineering
University of Glasgow

Dr R M Henry

Department of Chemical Engineering
University of Bradford

Mr A Lees

Engineering Research Station, British Gas
Newcastle-upon-Tyne

Professor J R Leigh

School of Electric Systems Engineering
Polytechnic of Central London

Professor D A Linkens

Department of Control Engineering
University of Sheffield

Professor D McLean

Department of Aeronautics
University of Southampton

Dr A S Morris

Department of Control Engineering
University of Sheffield

Dr N Mort

Department of Control Engineering
University of Sheffield

Dr D J Sandoz

Predictive Controls,
Manchester

Mr L S Smith

Eurotherm, Worthing

Mr N K Stanley

I B M, Greenock

Dr J D F Wilkie

British Sugar, Peterborough

Preface

This book is based on the lectures given at a Vacation School for postgraduate students in the areas of Control and Instrumentation held at the University of Sheffield in March 1990. For many years the Information Technology Directorate of the Science and Engineering Research Council has sponsored Vacation Schools, and this is the fourth school on the topic of computer control to be arranged by the Department of Control Engineering, University of Sheffield. Previous schools held in 1980, 1984 and 1987 led to the publication of books entitled *Computer Control of Industrial Processes*, *Real-Time Computer Control*, and *Computer Control*. The continually changing nature of the subject has led to the courses being re-structured on each occasion to reflect the current state-of-the-art, trends and possible future directions in the field. The Organizing Committee, Professor D.P. Atherton, Mr. C. Clark, Mr. P.C. Hammond, Professor D.A. Linkens, Dr. J.D.F. Wilkie played a major role in the planning of the course structure and we are grateful for their assistance. We also thank Mr. J.C. Leonard of the Science and Engineering Research Council for his advice and assistance.

The course had four major themes: design and tuning of controllers; the hardware technology; software design; and examples of applications. The first of the themes - design and tuning of controllers - is covered in chapters 1 -4. In chapter 1 Leigh covers discrete controller design for single-loop systems and in chapter 2 Virk deals with the design of controllers for multivariable systems. Methods of automatic tuning for commercial PID controllers are surveyed by Gawthrop chapter 3 and in chapter 4 practical aspects of implementing and tuning PID controllers are discussed by Smith.

The second theme - hardware technology - is introduced by Henry with a discussion of PLCs and their applications (chapter 5); Barney in chapter 6 covers networking technology for distributed computer control systems; and Virk deals with the technology of parallel processors for computer control applications (chapter 7).

Software design is introduced by Bennett in chapter 8 with a discussion of the particular problems of designing software for real-time control. In chapter 9, Mort describes one particular methodology, MASCOT, for the development of real-time software. Techniques for introducing fault tolerance into real-time software are described by Bennett in chapter 10. And in chapter 11 an example of the application of an object oriented approach to software design is described by Stanley.

The applications of computer control covered include control of robots (Morris, chapter 14), patient care including computer control of intensive care, anaesthesia and drug therapy (Linkens, chapter 13); batch control of sugar refining processes (Wilkie, chapter 16); modelling simulation and control of liquid gas vaporisers (Lees, chapter 12) and active control of fighter aircraft (McLean). Participants on the course also had the benefit of visits to ICI Huddersfield and to BSC Stainless, Sheffield where they were able to see applications of computer control: we thank the two companies for arranging the visits.

We are grateful for the support received from the staff and students of the Department of Control Engineering, University of Sheffield and thank them and Professor D.A. Linkens for making available the facilities of the Department for the course.   Particular thanks are due to Mrs. Margaret Vickers and Mr. R.D. Cotterill for administrative support.   We also thank all the lecturers who contributed to the course and to this book.   The preparation of the book was well supported by Mr. J.D. StAubyn of Peter Peregrinus and we thank him for his understanding and patience.   Finally we acknowledge the financial support of the Science and Engineering Research Council that made running the course possible.

Dr. S. Bennett
Dr. G.S. Virk

Department of Control Engineering
University of Sheffield

Chapter 1

Discrete controller design

J. R. Leigh

1) Introduction

The title, being interpreted, means the theory and practice of discrete time control (ie. the sort that is implementable in computers) to achieve closed loop control of continuous time processes. Thus, we are concerned with the discrete time control of continuous time processes within a hybrid feedback loop, figure 1. A natural initial question suggests itself: what are the advantages of discrete, as opposed to continuous, controllers? The answer is very disappointing. As far as control theory is concerned, there are no advantages. (Proof: every realisable discrete time signal is a continuous time signal but the converse is false). The reason for studying discrete time control is essentially practical: to allow reliable miniature low-cost digital electronic devices to be used as controllers.

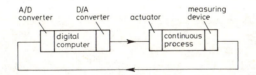

Fig. 1   A continuous process under closed-loop digital control

2)   Approaches to algorithm design

Roughly, there are two approaches:

2.1   Direct Controller Synthesis

Procedure in outline:

i)   Convert the specification that the final system must meet into a desired transfer function H(z).

ii)   Produce a transfer function G(s) representing the process that is to be controlled.

iii)   Form the transfer function $G'(s) = G_o(s)G(s)$, where G_o is a model of the interface between controller and process.

iv)   Discretise the transfer function G'(s) to produce the discrete time equivalent G'(z).

v)   Use the relation $D(z) = H(z)/\{G'(z)[1 - H(z)]\}$ (1) to synthesise the necessary controller for insertion into the loop (see figure 2).

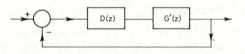

Fig 2. Synthesis in the z-domain

vi)   Convert D(z) into a difference equation and use it as a real time algorithm.

Comment

It can be seen that equation 1 contains models both of the process and the desired behaviour. In effect, the controller cancels out the existing process characteristics and replaces them by those of the required system.

2.2   Gain plus compensation approach

Idea in outline

i)   If a controller consisting of only a simple gain of numerical value C is used as in figure (3) then the performance of the resulting system (of transfer function $CG(z)/[1 + CG(z)]$) may be manipulated by choice of the value for C.

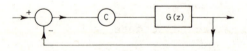

Fig 3. A controller consisting of a simple gain C

ii) As C is increased, the speed of response of the
     system increases but in general the response becomes
     oscillatory and as C is increased further, the system
     becomes unstable.

iii) By incorporating a suitable compensator M into the
     loop (figure 4) 'improved stability characteristics'
     can be given to the loop and then the value of C can
     be further increased with a consequent increase in
     speed of response. This process of juggling the
     design of compensator M and the value of gain C can
     be iterated until a best possible response is
     achieved.

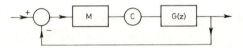

Fig 4. Incorporation of a compensator M into
       the loop

3)   Discussion of the design approaches

3.1   Direct Controller Synthesis

i)   Conversion of the specification into a desired
     transfer function H(z)

This step will very often involve a considerable amount of
approximation - particularly in those frequently
encountered cases where the original specification is
expressed in terms far removed from those pertaining to
transfer functions.

However, if the specification can be expressed in terms of
a desired natural frequency and a desired damping factor
then figure 5 may be used directly to choose the poles of
H(z).

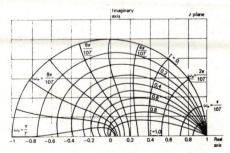

Fig 5. Diagram to assist in choosing the poles of
       H(z)

To use figure 5 decide natural frequency ω_n, damping factor ς, sampling interval T, and use the diagram to locate the intersection in the complex plane of the ω_n/T and the ς loci. Suppose this intersection is at a + jb, then the poles of the sought-for transfer function H(z) have to be located at a ± jb. That is, the denominator of H(z) should be (z - a + jb)(z - a - jb). The choice of the numerator of H(z) will be discussed later.

We note that most satisfactory control systems have damping factors in the 0.7 to 0.8 region. Thus the designer has, in general, little choice in regard to damping factor. The main task for the system designer is therefore to choose natural frequency.

Some guidelines to assist the choice of natural frequency are

Bandwidth $\quad \omega_b \simeq \omega_n$ $\qquad$ (2)
Rise time $\quad = \omega_n/2\,\varsigma$ $\qquad$ (3)

These relations apply to systems with a dominant pair of complex poles - see Leigh 1987).

Equation 2 links natural frequencies to closed loop bandwidth and through to the highest command frequency that can be followed and/or the highest disturbance frequency that can be actively rejected (ie. be rejected by the controller synthesising equal and opposite signals to those of the disturbance).

Equation 3 links natural frequency to rise time and settling time (assuming usual damping factor values are to be implemented) of closed loop step response - thus allowing a required closed loop step response to be translated from the specification through into a required natural frequency.

ii)   Production of G(s) - (process modelling)

In simple cases (rarely encountered in practice) the process is described by a known time-invariant differential equation, or transfer function G(s) with known coefficients.

Identifying G(s) from input/output data

Alternatively, the classical method of identifying G(s) is to use a frequency response analyser or, for noisy processes, a cross-correlator, connected directly to the process. More complex processes have to be modelled by setting up a nonlinear model whose structure is decided by theoretical considerations and whose parameters are chosen by numerical hill climbing searches on logged data. Linearisation and model reduction then produce a transfer function G(s) for inclusion in the control design processes (see Leigh {1983}).

Identifying G(z) from input/output data

An alternative approach to obtaining G(z) is to note that

> G(z) = y(z)/u(z) and letting G(z) = P(z)/Q(z) where P(z),Q(z) are polynomials in z (or equivalently in z^{-1} we obtain:

$$Q(z)y(z) = P(z)u(z) \qquad (4)$$

let n be the order of G(z) then equation 4 contains at most 2n + 2 unknown coefficients and therefore if we log 2n + 2 input/output pairs u_k, y_k, k = 1,..., 2n + 2, there will be sufficient data to calculate the unknown coefficients, for instance by matrix inversion or if more than 2n + 2 pairs are obtained, by pseudo - inversion, leading to a least squares fit.

iii)   Forming the transfer function G'(s). Suppose that a discrete time controller produces an output sequence [u(k)] that is intended to control a continuous process of transfer function G(s). Usually, the sequence will be converted into a piecewise constant analogue signal by an analogue to digital (A/D) converter before being applied to G(s). The mathematical equivalent of an A/D converter is a zero-order hold of transfer function $G_o(s)$ = {(1 - exp[-sT]}/s).

Note carefully that $Z\{G_1(s) \, G_2(s)\} \neq G_1(z)G_2(z)$ and therefore it is necessary to work with a composite transfer function G'(s) = $G_o(s)G(s)$ that represents the merging of the A/D converter and the process. This strategy is in its way a little masterpiece: distortion introduced by the very crude reconstruction by the A/D converter becomes part of the process that is to be controlled.

iv)   Discretisation of G'(s) to produce G'(z)

G'(s) may be discretised by any one of several alternative methods to produce G'(z). (But note that crude discretisation approaches introduce serious errors so that G'(z) will not be a true equivalent of the G'(s) that was the starting point. In fact it is quite possible for discretisation to produce an unstable G'(z) as the supposed equivalent of a strongly stable G'(s) - see Leigh (1985) for discretisation methods and a discussion of the fidelity of discretisation).

v)   Conversion of D(z) to a difference equation/real time algorithm.

This is a simple operation except that under certain conditions, the difference equation may not be physically realisable as a real time algorithm. This arises when future values of input are called for by the algorithm. The remedy to this problem is to return to H(z) and to modify its numerator. At device level, D(z) can be realised by several different layouts all of which result in the same mathematical operation but which may differ markedly in speed and performance with regard to rounding errors (see Leigh {1985}).

vi)   Choice of the numerator of H(z)

In choosing the numerator of H(z) the following factors need to be considered:

a) steady state response
b) frequency response
c) physical realisability and computational
   time requirements for D(z)

Considering a), recall that the steady response to a unit step, for stable H, is H(z) as z $\longrightarrow$ 1. Considering b), one point of view is that the response of H(z) when $\omega = \omega_s/2$ should be zero. Such behaviour can be obtained by placing one or more zeros at z = $^-$1. Considering c), notice that if the order of numerator and denominator of D(z) are equal then 'instantaneous' calculation of control outputs is implied. Choosing the order of numerator to be one less than the order of the denominator allows one sampling period T for the control algorithm to be calculated.

3.2.   Gain plus compensator approach

i) ii)   The root locus diagram is the classic tool for choosing the value for C. Figure 6 shows how, for a second order system, the pole locations are related to damping factor and natural frequency of the closed loop system.

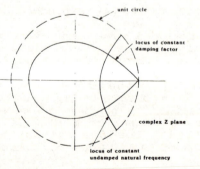

Fig 6. Desirable locations for the poles of a
       second order system may be chosen with
       the aid of the diagram

iii)   The compensator M primarily needs to improve the stability margin of the loop hence allowing higher gain C to be used, resulting in faster response. M may be an approximate differentiator, as in the <u>three term controller</u> (the three parallel terms are again C, differentiator M and an integrator I that is present to remove steady state error). Three term controllers are favoured by practitioners on grounds of: one form of controller satisfies all applications; the controller is easily 'tuned' for application using Ziegler - Nichols rules (Leigh {1987}); the controller is a successful work-horse being applied in huge numbers across industry.

      Seen from a frequency response point of view, M is a <u>phase-advance network</u> and frequency response techniques, usually used in the s domain, allow the design to be matched to the application.

4)   <u>Choice of sampling interval T</u>

      Shannon's theorem tells that a signal f(t) may be perfectly reconstructed from sampled data f* provided that the sampling interval T satisfies

$$T < \pi/2\omega_b$$

where ω_b is the highest frequency present in the signal f(t).

However, the algorithm to achieve this reconstruction uses the infinite data set f*, from $-\infty$ to $+\infty$ and assumes that the spectrum of f(t) is cut off vertically at ω_b.

In practice, reconstruction must rely on just a few (usually two) values of f*, with future values being unavailable in real time application. Additionally, cut off at ω_b is never vertical so that there are always frequencies greater ω_b than present in the signal.

These departures from the ideal mean that a shorter sampling interval than described above needs to be used. The recommended choice is

$$T < \pi/10\omega_b$$

Here ω_b is the closed loop bandwidth that is already part of the design specification or, if not, that can be approximately calculated from the required specification.

Note that if a larger sampling interval T has to be used. For instance:

$$T < \pi/5\omega_b$$

then experience shows that such a system will be usable but it will have a longer settling time than a system with the shorter sampling interval. Thus, the effect of using too long a sampling interval is first an increase in settling time then oscillations and only in extreme cases instability.

It is recommended, as an exercise, to plot a root locus for a closed loop system with all parameters except T fixed. Such a diagram is an aid to understanding the effects of varying T.

5)   Special Algorithms

5.1   Dead beat control

Let H(z) be an nth order transfer function all of whose poles are at the origin. When such a transfer function is subjected to a step input, its output will move by the end of n sampling intervals T to its final value where it will then remain, (figure 7). Control that aims to achieve this type of behaviour is called <u>dead beat control</u>.

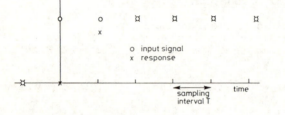

Fig 7. Step response of a second order transfer function G(z) having both poles at z = 0

5.2)   Takahashi's algorithm

In representing a typical process response by discrete data points (assuming that a constant value of sampling interval T is to be used), in order to capture the all important initial curvature, a rather short value of T is indicated. However, in order to capture the (also important) final value, a large value of T is indicated - so that the number of points to be logged will not be excessive.

Takahashi solves this problem nicely by taking frequent samples initially in the step response and then using a formula to generate further points until the correct steady state is reached (figure 8).

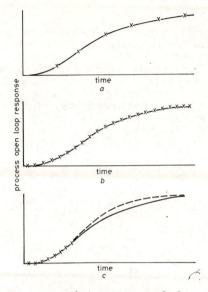

Fig 8. How many points are needed to capture a
        step response?

      a)   Two few points fail to capture the
           essential shape

      b)   Too many points to handle

      c)   Takahashi's approach. Early points
           capture the essential shape.
           Approximation (shown dotted) completes
           the response

Takahashi's algorithm then uses the model to
synthesise a controller for the process (that
generated the open loop step response) as follows
(figure 9).

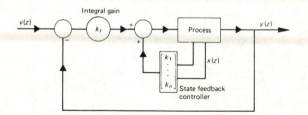

Fig 9. Takahashi's algorithm

5.3   Adaptive control

For control of a time-varying process, a time varying controller may be required.   There are very many approaches.   The most obvious approach is to produce an up-to-date process model using a recursive version of that outlined earlier, under the title: Identifying G(z) from input/output data (section 3.1), followed by choice of controller parameters to ensure satisfactory pole placement for the closed loop system.   The general layout for adaptive control is shown in figure 10.

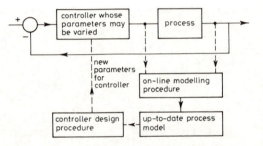

Fig 10. The general approach to adaptive control of a time-varying process

5.4   Predictive iterative control

Certain classes of difficult control problems have been solved as follows (figure11).

A set of possible control strategies is rapidly and repeatedly applied to a fast model of the process to be controlled and iteration is continued until evaluation shows that a worthwhile strategy has been found.   This strategy is then automatically implemented in the real controller and the search for the next strategy begins again using the fast model.

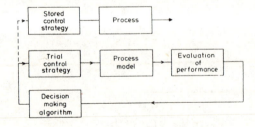

Fig 11. Predictive-iterative control strategy

6)   <u>Limits to attainable performance</u>

The fastest rate at which a particular water tank can be filled through a particular pipe depends on the maximum pressure available to drive water through the pipe. This obvious statement is included as a reminder that performance is nearly always constrained by process factors and that these factors represent nonlinearities - absent by definition from our linear models, ie the methods of control system synthesis do not contain any constraints or mechanisms for taking constraints into account.

This means that a control system must be designed entirely ignoring constraints. However when such a system is built, the constraints will make their presence felt and the system will not perform as well as expected.

For instance, let an electric motor have the transfer function $G(s)$. The transfer function contains no information on the maximum voltage, maximum current, maximum safe shaft power or on the duty limitations caused by heating. Yet the transfer function and the required performance are the only inputs into the design process.
Physical upper limits on mechanical, electrical, thermal, hydraulic etc. variables form one group of factors that constrain performance. These can be roughly categorized as <u>actuation limits</u>.
The second set of limiting factors are <u>measurement problems</u>. These are caused by lack of resolution of sensors, drift and error in measurement and generally, limitations to the system caused by measurement limitations.

Finally, there are <u>noise problems</u>. After very high amplification all practical signals appear noisy and eventually this effect will limit attainable performance.

7)   <u>Overview: Concluding comments, guidelines for algorithm choice and some comments on procedure</u>

i)   Very broadly, there are two approaches to algorithm design. The first, synthesis of $D(z)$ to achieve a specific closed loop transfer function $H(z)$ is theoretically sound but suffers from two defects: choosing $H(z)$ usually involves massive approximation; $D(z)$ 'contains' both $G(z)$ and $H(z)$ and is therefore often unwieldy. The second approach using a gain plus compensator, is not very scientific but it has the great merit of simplicity.

ii)   Every continuous time algorithm can be discretised - this is one source of algorithms. Note however that the performance of a discretised algorithm is always degraded to some extent, compared with that of the original. The extent of degradation is governed by the choice of sampling interval.
These are however discrete time algorithms that have no (apparent) continuous time equivalents. These are the most interesting algorithms and they tend to be incorporated as part of advanced control packages for solution of demanding problems.

iii)   Some industries, like aerospace, tend predominantly to use frequency response continuous time design methods and only later to discretise.
Process industries tend to use off-the-shelf three term algorithms integrated within diagnostic and monitoring supervisory software.

iv)   In general, it is recommended to use simple solutions (for instance off-the-shelf three term controllers) for routine problems.
However, it is important to match the level of sophistication of the controller to the inherent difficulty of the problem.

(Too often one encounters inadequate ad-hoc solutions that have been built piece-wise with no support from available control methods).

v)   Many alternative methods have been put forward for the selection of sampling interval T. The one suggested here, based on closed loop bandwidth is a reasonable compromise between ad-hoc methods and theoretical overkill.

8)   References and further reading

        The books by Kuo and Ogata are large and comprehensive. The small book by Katz has a rather staccato style - sometimes it is like a designer's rough note book - but this aspect makes it valuable. The older book by Takahashi is one of the author's favourites. It contains many authoritative and realistic examples.

        For identification, {of G(s), G(z)} the recent book by Soderstrom is highly recommended. The book, through theoretically strong, nevertheless deals with all necessary practical details.

        Illustrative examples of particular applications may be found by scanning the contents list of the last ten years of the journal 'Automatica'. Leigh(1985) and Leigh(1987) both contain illustrative application examples as well as bibliographies containing papers describing applications.

Franklin, G F and Powell, J D., 1980, Digital Control, Addison-Wesley, Reading, Mass

Houpis, C H and Lamont, G B., 1985, Digital Control Systems, McGraw-Hill Book Company, New York

Katz P., 1981, Digital Control using Microcomputers, Prentice-Hall International, London

Kuo B C., 1980, Digital Control Systems, Holt Rinehart and Winston, New York

Leigh, J R., 1983, Modelling and Simulation, IEE Topics in Control Series No 1, Peter Peregrinus Ltd, Stevenage, UK

Leigh J R., 1985, Applied Digital Control; Theory, Design and Implementation, Prentice-Hall International, Englewood Cliffs, New Jersey

Leigh J R., 1981, Applied Control Theory, IEE Control Engineering Series No 18, Peter Peregrinus Ltd, Stevenage, UK

Ogata K., 1987, Discrete-time Control Systems, Prentice-Hall International, London

Soderstrom T and Stoica P., 1989, System Identification, Prentice-Hall International, New York

Takahashi Y , Rabins M J and Auslander D M., 1970, Control and Dynamic Systems, Addison-Wesley, Reading, Mass

Chapter 2

Multivariable control system design

G. S. Virk

1 Introduction

In the analysis and design of single input-single output (SISO) systems (as shown in Fig 1(a) a transfer function is usually the starting point. This can be analysed directly using the classical methods of Bode, Nyquist, etc, or it can be represented in state-space form for which, the so called modern methods are available. When dealing with multivariable systems the same general philosophy can be applied and extensions of the methods applied.

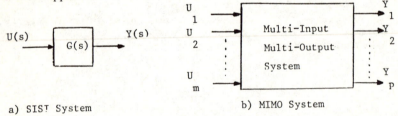

a) SIS^{T} System   b) MIMO System

Figure 1: Open-Loop Control Systems

Consider for example a system with m inputs $U_1, U_2, \ldots, U_m$, and p output $Y_1, Y_2, \ldots, Y_p$, as shown in Fig 1(b). Such a system can be represented in transfer function form as:

$$Y(s) = G(s)U(s) \tag{1}$$

$$where \quad G(s) = \begin{bmatrix} G_{11}(s) & G_{12}(s) & \ldots & G_{1m}(s) \\ G_{21}(s) & G_{22}(s) & \ldots & G_{2m}(s) \\ \vdots & \vdots & & \vdots \\ G_{p1}(s) & G_{p2}(s) & \ldots & G_{pm}(s) \end{bmatrix}$$

and $Y=[Y_1, Y_2, \ldots, Y_p]^T$, $U=[U_1, U_2, \ldots, U_m]^T$, and a^T represents the transpose of a vector, and each element $G_{ij}(s)$ is the transfer function relating input j with output i, etc. This can be converted, using standard programming techniques to give the standard state-space representation:

$$\dot{x}(t) = Ax(t) + Bu(t)$$
$$y(t) = Cx(t)$$

where $\quad$ A $\quad=\quad$ n×n system matrix (where n is the order of the system).

$\qquad$ B $\quad=\quad$ n×m input matrix.

$\qquad$ C $\quad=\quad$ p×n output matrix.

We will restrict attention, in the main, to strictly proper closed-loop systems having m inputs and m outputs, and unity negative feedback. Considering such a system and inserting a proper forward-path controller K(s) as shown in Fig 2, we can analyse the behaviour of the closed-loop system using similar methods applicable to SISO systems.

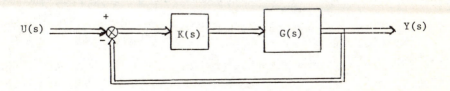

Figure 2: Closed-Loop Compensated MIMO System

In some aspects the same methods apply in both cases. For example the closed-loop transfer function matrix has the form

$$H_c\left(s\right) = \left(I_m + Q\left(s\right)\right)^{-1}Q\left(s\right) \tag{2}$$

where Q(s) = G(s)K(s) and I_m is the unity matrix of dimension m. The stability of this can be assessed by first forming the characteristic polynomial

$$\frac{\rho_c\left(s\right)}{\rho_o\left(s\right)} \; = \; |\,T\left(s\right)\,| \; = \; |\,I_m + Q\left(s\right)\,| \tag{3}$$

and then applying the usual Routh technique to the numerator of $|\,T(s)\,|$, where $\rho_c(s)$ and $\rho_o(s)$ are the closed-loop and open-loop characteristic polynomials, respectively. The following result from Owens [3], shows how the Nyquist method is applied to MIMO systems.

1.1   Theorem: Multivariable Nyquist Stability Criterion

Let D be the usual Nyquist contour in the complex plane, assume $\rho_c(s)$ and $\rho_o(s)$ have n_c and n_o zeros in the interior of D respectively, and that Γ is the closed contour generated by T(s) as s varies on D in a clockwise manner. Then

$$n_c - n_o = n_T \tag{4}$$

where n_T is the number of clockwise encirclements of Γ about the origin, and the closed-loop system is asymptotically stable if and only if

$$n_c = 0, \;\; i.e, \;\; n_o + n_T = 0 \tag{5}$$

Other concepts we shall require in the subsequent analysis are that of diagonal dominance and Gershgorin's circles.

1.2   Definition

An n×n matrix A is said to be row (column) dominant if

$$| a_{ii} | > \sum_{j=1, j \neq i}^{n} | a_{ij} | \qquad \left(\sum_{j=1, j \neq i}^{n} | a_{ji} | \right) \qquad (6)$$

for i = 1, 2, ..., n, where a_{ij} are the elements of A.

1.3   Definition

Let A be an n×n complex matrix and plot the n diagonal terms a_{ii}, $1 \leq i \leq n$, as points in the complex plane. At each point a_{ii}, plot a circle of centre a_{ii} and radius

$$r_i = \sum_{j=1, j \neq i}^{n} | a_{ij} | \qquad (row \;\; estimate). \qquad (7)$$

Each such circle is called a Gershgorin circle. Using these we have the following result (Owens [3]):

1.4   Theorem

Let $I_m + Q(s)$ be diagonally - (row or column) dominant at every s on the D contour. Let the diagonal transfer functions $q_{ii}(s)$ map D onto closed contours C_{ii} encircling the (-1, 0) point of the complex plane n_i times, $1 \leq i \leq m$, in a clockwise manner. Then

$$n_T = \sum_{i=1}^{m} n_i \qquad (8)$$

and the closed-loop system is asymptotically stable if and only if

$$n_o + \sum_{i=1}^{m} n_i = 0$$

What this result states is that under diagonal dominance the stability of the m input/m output system can be investigated by the anlysis of m SISO systems with forward-path transfer functions $q_{ii}(s)$, $1 \le i \le m$. Hence, in multivariable systems the obvious aim for designing compensators is to ensure that $Q(s) = G(s)K(s)$ is diagonally dominant on D. This unfortunately is not a trivial problem. The existence of a suitable $K(s)$ is guaranteed if $G(s)$ is invertible, in which case we can set

$$K(s) = G^{-1}(s)diag\left\{q_{ii}(s)\right\}_{1 \le i \le m} \tag{9}$$

so that

$$I_m + Q(s) = diag\left\{1 + q_{ii}(s)\right\}_{1 \le i \le m} \tag{10}$$

This choice of $K(s)$ results in a non-interacting controller which suffers from stability and non-minimum phase problems, together with being of a complex structure. For practical purposes we require a controller having some simple form. Achieving such a design is not obvious although various algorithmic approaches exists to give a suitable $K(s)$ for stability purposes. These methods give little insight into the closed-loop transient behaviour, and a more suitable approach is provided by the Inverse Nyquist Array method proposed by Rosenbrock [6] .

2   Inverse Nyquist Array

Using the notation $\tilde{L}(s) = L^{-1}(s)$ to represent the inverse of any invertible transfer function matrix $L(s)$ we have that the inverse of the closed-loop transfer function matrix takes the form

$$\tilde{H}_c(s) = I_m + \tilde{Q}(s) \tag{11}$$

This provides a useful link between the inverse forward-path system $Q(s)$ and the closed-loop dynamics. The relation with the closed-loop stability can be obtained by looking at

$$| T(s) | = | I_m + Q(s) | = \frac{| I_m + \tilde{Q}(s) |}{| \tilde{Q}(s) |} \tag{12}$$

We also have the following result:

2.1   Theorem

Let $I_m + \tilde{Q}(s)$ and $\tilde{Q}(s)$ be diagonally - (row or column) dominant on D. Let the diagonal elements $\tilde{q}_{ii}(s)$, $1 \leq i \leq m$, of $\tilde{Q}(s)$ map D onto closed contours in the complex plane encircling the origin and the point (-1, 0), $\tilde{n}_i$ and $\hat{n}_i$ times respectively, $1 \leq i \leq m$, in a clockwise manner. Then the closed-loop system is asymptotically stable if and only if

$$n_o + \sum_{i=1}^{m} (\hat{n}_i - \tilde{n}_i) = 0 \tag{13}$$

Here again we can use the Nyquist method to assess the stability of the overall system by looking at the diagonal terms of $\tilde{Q}(s)$ on D together with the Gershgorin circles. The advantage with using the inverse system rather than the original is that the dominance conditions can be achieved using simple controllers (see below). These techniques can be extended to provide information on more accurate estimates of closed-loop behaviour by the use of Ostrowski's theorem, presented next.

2.2   Theorem

Let A be a complex m×m matrix which is diagonally-row (column) dominant. Then

$$| (\tilde{a}_{ii})^{-1} - a_{ii} | \leq \left(\sum_{k=1, k \neq i}^{m} a_{ik} \right) max_{1 \leq j \leq m, j \neq i} \frac{\sum_{k=1, k \neq j}^{m} | a_{jk} |}{| a_{jj} |} \tag{14}$$

for i = 1, 2, ..., m. Applying this to $\tilde{H}_c$ (s) (assumed row dominant) we have

$$| (h_{ii})^{-1} - \tilde{h}_{ii} | \leq \left(\sum_{k=1, k \neq i}^{m} | \tilde{h}_{ik}(s) | \right) max_{1 \leq j \leq m, j \neq i} \frac{\sum_{k=1, k \neq j}^{m} | \tilde{h}_{jk}(s) |}{| \tilde{h}_{jj} |} \tag{15}$$

This result has the following graphical interpretation:

Suppose the plots of $\tilde{q}_{ii}(s)$, $1 \leq i \leq m$, on D are drawn with superimposed Gershgorin circles of radius $\tilde{r}_i(s)$ based on row or column estimates. If the critical (-1, 0) point does not lie in or on any such circle, then $I + \tilde{Q}(s)$ is dominant on D. At each point on D calculate the radii

$$d_i(s) = max_{1 \leq j \leq m, j \neq i} \frac{r_j(s)\, r_i(s)}{1 + \tilde{q}_{ij}(s)} \qquad i = 1, 2, \ldots, m \qquad (16)$$

and plot the diagonal terms $\tilde{q}_{ii}(s)$ on D, i=1,2,...,m with superimposed circles of radius $d_i(s)$. As s goes round D these circles sweep out bands lying inside the Gershgorin bands. These bands are called Ostrowski bands. If they are narrow enough, then the closed-loop response can be assessed using the approximation

$$h_{ii}^{-1} \approx 1 + \tilde{q}_{ii}, \qquad 1 \leq i \leq m, \quad s \; on \; D, \qquad (17)$$

assuming that interaction effects are negligible.

2.3   Achieving Dominence

It is possible to design a suitable K(s) by trial and error but for m≥3 this can be very difficult or nearly impossible and a more systematic approach is needed (see Rosenbrock [6]). We illustrate the approach by an example.

2.3.1   Example

$$If \; G(s) = \frac{1}{3s + 2} \begin{bmatrix} -s & s+1 \\ s+2 & -s \end{bmatrix} \implies \tilde{G}(s) = \begin{bmatrix} s & s+1 \\ s+2 & s \end{bmatrix}$$

This G(s) is clearly not dominant and we need to add a compensator to operate on its rows.

$$Choosing \quad \tilde{K} = \begin{bmatrix} 0 & 1 \\ 1 & 0 \end{bmatrix}, \quad gives \quad \tilde{K}\tilde{G}(s) = \begin{bmatrix} s+2 & s \\ s & s+1 \end{bmatrix}$$

which is both row and column dominant. Therefore the controller needed is

$$K = \begin{bmatrix} 0 & 1 \\ 1 & 0 \end{bmatrix}$$

In most cases it is not practical to make Q(s) diagonally dominant at all frequencies but at only one. This frequency is usually chosen to be $\omega = 0$ because G(0) is real and so a constant controller is adequate. For further discussion on this aspect of controller design see Rosenbrock [6], [7], and Owens [3].

3 Multivariable Root-Loci

The classical root-locus method has been a valuable tool in the analysis of control systems. In this section we extend the technique to multivariable systems. An m input/m output strictly proper system G(s) with unity negative feedback and forward-path controller K(s) has the following relation from above

$$\frac{\rho_c(s)}{\rho_o(s)} = |T(s)| = |I_m + Q(s)| \tag{18}$$

where Q(s) = G(s)K(s). The classic root-locus technique gives the variation of the closed-loop poles as the controller gain is varied. In the present multivariable setting we have m^2 gains that can be varied in any independent manner giving rise to many different root-loci. To simplify the complexity we restrict attention to controllers of the form $K(s) = pK_1(s)$, where $K_1(s)$ is a specified controller and p> 0 is a real scalar that represents the overall gain. The root-locus plot of the closed-loop system is then defined to be the graphical representation of the variation of the roots of $\rho_c(\mathrm{s})$ as p varies in the interval $0 \leq p < \infty$. It is generally not possible to obtain analytic expressions for the closed-loop poles and so some form of approximation has to be made. As in the SISO case the root-loci start at open-loop poles and terminate at zeros of Q(s) - some of these being at infinity. One infinite zero, in a system implying the root-locus goes to infinity along the negative real axis, two implying $\pm 90°$ asymptotes, etc. We therefore have for the MIMO case that the root-locus has finite cluster points at the zeros of Q(s), and a variety of unbounded poles of the form

$$s_{ij} \;=\; p^{\frac{1}{\nu_i}} \eta_{ij} + \alpha_i + \varepsilon_{ij}(p) \tag{19}$$

$$lim_{p \to \infty} \;\; \varepsilon_{ij}(p) \;=\; 0, \qquad 1 \leq j \leq \nu_i, \quad 1 \leq i \leq m. \tag{20}$$

where η_{ij} are the distinct ν_i^{th} roots of a non-zero number $-\lambda_i$. Relation (19) is said to represent an infinite zero of order ν_i with asymptotic directions η_{ij} and pivot α_i.

3.1   Uniform Rank Systems

The calculation of the orders, asymptotic directions and pivots of the root-locus is undertaken by analysis of the following series about the point at infinity

$$Q(s) = s^{-1}Q_1 + s^{-2}Q_2 + s^{-3}Q_3 + \cdots \tag{21}$$

where Q_i, i=1,2,... are the system Markov parameters and are given in terms of the state-space matrices by $Q_i = CA^{i-1}B$, $i \geq 1$. A special case of interest is when $Q_i = 0$, for $1 \leq i < k$, $|Q_k| \neq 0$, in which case the system is said to have uniform rank k. Suppose that s is a closed-loop pole and that $|s| \to \infty$ as $p \to \infty$ then

$$0 = |T(s)| = |I_m + \frac{p}{s^k}Q_k + \frac{p}{s^{k+1}}Q_{k+1} + O\left(\frac{p}{s^{k+2}}\right)| \tag{22}$$

where the notation $O\left(\frac{p}{s^l}\right)$ is used to represent the fact that $p^{-1}s^l O\left(\frac{p}{s^l}\right)$ remains finite as $p \to \infty$. Multiplying equation (22) by $\left(p^{-1}s^k\right)^m$ yields

$$0 = \mid \frac{s^k}{p}I_m + Q_k + O\left(s^{-1}\right) \mid \qquad (23)$$

Letting $p \to \infty$, the term $O\left(s^{-1}\right) \to 0$ and $p^{-1}s^k$ must have only finite cluster points λ satisfying

$$\mid \lambda I_m + Q_k \mid = 0 \qquad (24)$$

that is $Q(s)$ has mk k^{th} order infinite zeros of the form

$$lim_{p\to\infty} \; \frac{s^k}{p} = -\lambda_i \qquad (25)$$

where $\lambda_i \neq 0$, $1 \leq i \leq m$ are the eigenvalues of Q_k. Equation (25) imples that the infinite zeros take the form

$$s = p^{\frac{i}{k}}\eta_{ij} + \mu_{ij}(p), \qquad 1 \leq i \leq m, \qquad 1 \leq j \leq k \qquad (26)$$

where $\eta_{ij}, 1 \leq j \leq k$ are the distinct k^{th} roots of $-\lambda_i$ and

$$lim_{p\to\infty} \; p^{\frac{-1}{k}}\mu_{ij}(p) = 0 \qquad (27)$$

The pivot can only be obtained by analysis of the $\mu_{ij}(p)$ terms. Suppose $\{\lambda_i\}$ are all distinct for i=1,2,...,m and that

$$M^{-1}Q_kM = diag\left\{\lambda_i\right\}_{1\leq i\leq m} \qquad (28)$$

and let

$$M^{-1}Q_{k+1}M = N = \begin{bmatrix} n_{11} & n_{12} & \cdots & n_{1m} \\ n_{21} & n_{22} & \cdots & n_{2m} \\ \vdots & & & \vdots \\ n_{m1} & n_{m2} & \cdots & n_{mm} \end{bmatrix}$$

Examining the identity

$$0 = | \frac{s^{k+1}}{p} M^{-1} T(s) M |$$ (29)

it can be shown that equation (26) can be written as

$$s(p) = p^{\frac{1}{k}} \eta_{ij} + \frac{n_{ii}}{p\lambda_i} + \varepsilon_{ij}(p)$$ (30)

and $lim_{p \to \infty}$ $\varepsilon_{ij}(p) = 0$, see Owens [3] . By considering the next Markov parameter

the rate of approach to the asymptotes may be obtained, see Owens and Virk [4]. Also

general non-uniform rank systems can be studied in this way, except that the infinite

zeros will be of different order, see Owens [3].

4   State-Space Methods

The m input/m output system considered in sections 1-3 can be represented in state-

space form as shown in Fig 3(a) to give the usual representation:

$$\dot{x}(t) \quad = \quad Ax(t) + Bu(t)$$ (31)

$$y(t) \quad = \quad Cx(t)$$ (32)

where x is the system state, A is an n×n system matrix, B is an n×m input matrix, C is

an m×n output matrix, and n is the order of the system. Depending upon the technique

used in obtaining the state-space representation, the matrices can have various forms

- for example, A may be in companion form or be diagonal. To revert back to the

transfer function form we have the following relation

$$G(s) = C \left[sI_n - A \right]^{-1} B$$ (33)

Equations (31) and (32) can be solved to yield ($t_0 = 0$):

$$x(t) = \Phi(t)x(0) + \int_0^t \Phi(t - \tau)Bu(\tau)d\tau \qquad and \;\; y(t) = Cx(t),$$ (34)

where $\Phi(t)$ is the state transition matrix and is given by either of the following methods:

1. $\Phi(t) = L^{-1}\left\{[sI_n - A]^{-1}\right\}$ where L^{-1} is the inverse Laplace transform operation.

2. $\Phi(t) = e^{At}$ where the matrix exponential e^{At} is defined by

$$e^{At} = I_n + At + \frac{1}{2!}A^2t^2 + \frac{1}{3!}A^3t^3 + \ldots \tag{35}$$

3. $\Phi(t) = Me^{\Lambda t}M^{-1}$ where $\Lambda = diag\ \{\lambda_i\}_{1\leq i\leq n}$ is the diagonal matrix whose elements are composed of the eigenvalues of the A matrix (assumed distinct) and M is the eigenvector matrix of A.

The state transition matrix is defined by the system matrix, A, and therefore defines the behaviour of the system.

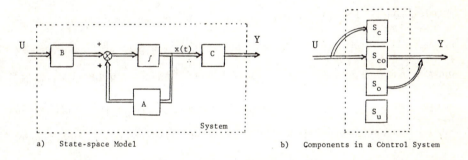

a)   State-space Model   b)   Components in a Control System

Figure 3: State-Space Block Representation

When considering design aspects in the state-space formulation, the following concepts are needed for assessing the goals that can be achieved, using for example, state feedback.

4.1   Definition: Controllability

A system is said to be state-controllable if it is possible to find a control function u(t) which will transfer the system state from any initial state $x(t_0)$ to any final state $x(t_1)$ in a finite time interval $t_0 \leq t \leq t_1$.

4.2   Definition: Observability

A system is said to be state-observable if measurements on the output y(t) contain sufficient information to completely determine the system state x(t).

Using these concepts, four possible subsections can exist in any system, namely, a section that is observable but uncontrollable (S_o), a section that is controllable and observable (S_{co}), a section that is controllable but not observable (S_c) and a section that is uncontrollable and unobservable (S_u). The subsections are shown more clearly in Fig 3(b). Out of the four it is only the S_{co} subsection that satisfies the transfer function relation Y(s) = G(s)U(s), and so is the only portion considered in the classical analysis methods. Hence the state-space approach gives more detail to the system representation and allows more insight into the internal workings of the overall system. Bad modes can be spotted and suitable controlling and/or outputting linkages can, therefore be designed and inserted.

Regarding controllability and observability we have the following result:

4.3   Theorem

Given a state-space representation as in equations (31) and (32), we then have

1. The system is state controllable if and only if the n×(n×m) matrix

$$\left[B \vdots AB \vdots A^2B \vdots \dots \vdots A^{n-1}B \right]$$ (36)

has rank n.

2. The system is observable if and only if the matrix

$$\left[C^T \vdots A^T C^T \vdots \left(A^T\right)^2 C^T \vdots \dots \vdots \left(A^T\right)^{n-1} C^T \right]$$ (37)

has rank n.

4.4 Linear State Variable Feedback

Controllability and observability tell us which modes in a control system can be controlled and/or observed. Looking at the control aspects, let is introduce feedback loops so that a closed-loop system is obtained. If these loops are made using the state vector we have the situation as shown in Fig 4(a). Hence we are using state feedback to give the controlling input as $u(t) = v(t) + Fx(t)$, where F is a m×n feedback matrix to be designed. In this case the state equation becomes

$$\dot{x}(t) = (A + BF)x(t) + Bv(t)$$ (38)

and so the response of the closed-loop system is governed by the n×n matrix $(A + BF)$ whose eigenvalues depend on the feedback loops F. By appropriate choice of F, and if the system is state controllable, we may alter the eigenvalues as required. The method is also known as pole-placement.

4.5 Output Feedback

If feedback compensation is applied to a state-space representation we get the following

Open-loop: $\dot{x} = Ax + Bu$

$y = Cx$

Closed-loop: the input now becomes v - Ky, as shown in Fig 4(b) and so

$\dot{x} = (A - BKC)x + Bv$

and so the closed-loop behaviour with output feedback is governed by the matrix $(A - BKC)$.

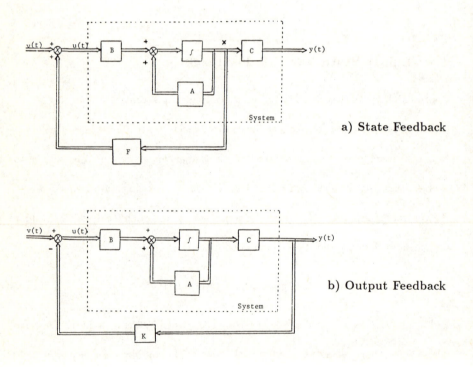

a) **State Feedback**

b) **Output Feedback**

Figure 4: Closed-Loop State-Space Systems

4.6   Observers

When attempting to use state feedback it is clear that the state vector needs to be accessible to enable the implementation of the feedback signals. Unfortunately this, in general, is not the case. We only have access to the input and output of the system under consideration, although, in certain cases, some state variables may be measurable. What is needed is a technique for the estimation of the state vector from the information available.

This can be done if A, B and C are known, by constructing a state observer (see Luenberger [2]), as shown in Fig 5.

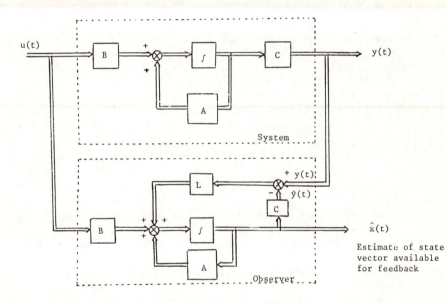

Figure 5: State Observer

The input and output from the system are used to drive the observer which, hence has a dynamical equation as

$$\dot{\hat{x}}(t) = (A - LC)\,\hat{x} + Ly(t) + Bu(t) \tag{39}$$

where $\hat{x}$ denotes the estimate of x, etc, and L is an n×m observer matrix.

Letting e be the error between the actual state vector and its estimate, that is

$e = x - \hat{x}$, then we can show that

$$\dot{e}(t) = \dot{x}(t) - \dot{\hat{x}}(t) = (A - LC)e(t) \tag{40}$$

Hence the error dynamics are defined by the matrix (A-LC) whose eigenvalues depend on the observer matrix L. If the system (31, 32) is observable then we can determine an L to arbitrarily assign the eigenvalues of (A-LC) and hence control the behaviour of the error as required.

If the eigenvalues of (A-LC) have negative real parts smaller than $-\sigma$, then all elements of e will approach zero at rates faster than $e^{-\sigma t}$ (note equation (40) has only a free motion solution). Therefore, even if the initial error $e(t_0)$ is large $\hat{x}$ will approach x rapidly. Once $\hat{x}(t) \to x(t)$, the two stay equal, and in addition, $\hat{x}$ is available for use in state-feedback, optimisation analysis, etc. In other words we can use $\hat{x}$ as if it were the true state vector. What must be noted is that we must design L so that (A-LC) has faster eigenvalues than the system states so that $e(t) \to 0$ rapidly. Also note that the whole technique relies on the fact that A, B, C are known to a high degree of accuracy. If this is not the case an adaptive form of observer needs to be considered.

The above analysis constructs what is called a full-order observer since the complete state vector $\hat{x}$ is constructed. In general the output $y \in R^p$ spans a subspace of the n-dimensional state-space and so contains information on some of the states. These are therefore directly available at the output of the system and hence do not need to be estimated. This gives rise to reduced-order observers, see for example, Chen [1].

5 Optimal Control Theory

When using optimal control theory to design controllers the performance is gauged by a performance index (or cost function). This may be maximised, or equivalently, minimised, depending on the formulation. For example we may wish to

1. Minimise energy (terms of the form u^2).

2. Minimise errors (terms of the form x^2).

3. Minimise time.

The class of problems that can be considered under this section is quite large and we do not have the space to go into details. We will therefore restrict attention to a special form of problem for which the optimal control law can be determined in a closed form. This is commonly referred to as the linear quadratic performance index (LQP) problem. This takes the following form for the continuous case:

$$Min_u \quad \frac{1}{2} \int_0^{t_f} \left[x^T(t)Qx(t) + u^T(t)Ru(t) \right] dt + \frac{1}{2} x^T(t_f)Qx(t_f) \quad (41)$$

$$subject \ to \quad \dot{x}(t) = Ax(t) + Bu(t) \quad (42)$$

$$x(0) = x_0 \quad (43)$$

and for the discrete case we have the form:

$$Min_u \quad V^u(x,0) = \frac{1}{2} \sum_{k=0}^{N-1} \left[x^T(k)Qx(k) + u^T(k)Ru(k) \right] + \frac{1}{2} x^T(N)Qx(N) \quad (44)$$

$$subject \ to \quad x(k+1) = Ax(k) + Bu(k) \quad (45)$$

$$x(0) = x_0 \quad (46)$$

The use of A, B and C to define the state-space representations in both continuous and discrete domains should not cause confusion. Concentrating on the discrete formulation, the notation used is that $V^u(x, 0)$ is the cost function for control u, starting at state x, and time $k = 0$. Q is an n×n matrix, R is an m×m matrix and A and B are matrices which define the system's state equations in difference form. We will need the following assumptions:

1. The matrix Q is positive semi-definite (written $Q \geq 0$) and symmetric.

2. The matrix R is positive-definite ($R > 0$) and symmetric.

We therefore need to define these concepts.

5.1   Definition

1. An r×r matrix M is said to be positive semi-definite if $x^T M x \geq 0$ for all vectors $x \neq 0$.

2. Similarly a matrix M is positive-definite if $x^T M x > 0$ for all vectors $x \neq 0$.

3. An r×r matrix M is symmetric if $M^T = M$.

We will make use of the following theorem:

5.2   Theorem

Suppose

$$J(u) = a + b^T u + \frac{1}{2} u^T C u \tag{47}$$

where C> 0 and symmetric, then J(u) has a unique minimum at u*, where

$$u^* = -C^{-1} b \tag{48}$$

and the corresponding minimum cost is

$$J(u^*) = a - \frac{1}{2}b^T C^{-1} b \qquad (49)$$

Essentially the problem as stated requires the determination of the control sequence u(0), u(1), u(2), ..., so that the quadratic objective is minimised. Hence N control values, or an N stage problem needs to be solved if the optimisation is to be done. If N is large, the amount of processing necessary to calculate the control sequence can be excessive. Fortunately for us, better programming techniques have been developed, and one such method is to use dynamic programming.

5.3   Dynamic Programming for Discrete LQP Problems

Dynamic programming reduces the N stage decision problem to N single stage subproblems that are related to each other via an iterative scheme. The following procedure is undertaken:

1. At k = N we know that

$$V^*(x, N) = \frac{1}{2}x^T Q x \qquad (50)$$

for all x, and Q≥ 0, symmetric where V* denotes the optimal value.

2. Assume at some stage k+1 we have

$$V^*(x, k+1) = \frac{1}{2}x^T P(k+1)x \qquad (51)$$

for all x where P(k+1)≥ 0 and symmetric.

3. Then the iterative relation we require is that

$$V^*(x, k) = \frac{1}{2}x^T P(k)x \qquad (52)$$

for all x and P(k) ≥ 0.

We can deduce relation (52) by considering the following "fundamental recurrence relationship" of dynamic programming:

$$V^*(x(k), k) = Min_{u(k)} \left\{ \frac{1}{2} x^T(k) Q x(k) + \frac{1}{2} u^T(k) R u(k) + V^* \left(x(k+1), k+1 \right) \right\} \quad (53)$$

Using this together with equation (51) and the dynamics of the system under consideration (45), we have

$$V^*(x(k), k) = Min_{u(k)} \left\{ \frac{1}{2} x^T Q x + \frac{1}{2} u^T R u + \frac{1}{2} (Ax + Bu)^T P(k+1) (Ax + Bu) \right\}$$

$$(54)$$

where all variables, unless specifically stated are evaluated at k. This can be rearranged to give

$$
\begin{aligned}
V^*(x(k), k) &= Min_{u(k)} \{ \frac{1}{2} x^T \left[Q + A^T P(k+1)A \right] x + x^T A^T P(k+1) Bu \\
&+ \frac{1}{2} u^T \left[R + B^T P(k+1)B \right] u \}
\end{aligned}
\quad (55)
$$

This is in the form to which Theorem 5.2 can be applied. Hence the minimising control u*(x,k) is given by

$$
\begin{aligned}
u^*(x, k) &= -\left[R + B^T P(k+1)B \right]^{-1} \left(x^T A^T P(k+1)B \right)^T \\
&= -\left[R + B^T P(k+1)B \right]^{-1} \left(B^T P(k+1)Ax \right) \quad (56) \\
&= -\Delta(k)x \quad (57)
\end{aligned}
$$

and the minimum cost $V^*(x, k)$ satisfies

$$
\begin{aligned}
V^*(x, k) &= \frac{1}{2} x^T \left[Q + A^T P(k+1)A \right] x \\
&+ \frac{1}{2} x^T A^T P(k+1)B \left[R + B^T P(k+1)B \right]^{-1} B^T P(k+1)Ax \quad (58)
\end{aligned}
$$

or $V^*(x, k+1) = \frac{1}{2} x^T P(k)x$ as required in equation (52), where we define P(k) by

$$P(k) = Q + A^T P(k+1)A + A^T P(k+1)B \left[R + B^T P(k+1)B \right]^{-1} B^T P(k+1)A \quad (59)$$

This P(k) can be shown to be positive semi-definite and symmetric.

Equation (59) is called the Riccati equation and needs to be solved backwards in time starting with the initial condition

$$P(N) = Q \qquad (60)$$

This matrix difference equation can be solved off-line so that the optimal control policy

$$u^*(x, k) = -\Delta(k)x \qquad (61)$$

for all x, all k can be determined and implemented in a feedback loop as shown in Fig 6.

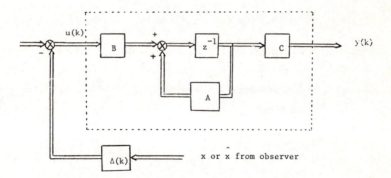

Figure 6: Optimal Control Implementation

6   Conclusion

An introduction to multivariable control system design has been presented. Due to lack of space only a brief overview has been given but the interested reader can consult the references cited for further discussion.

References

[1] Chen, C T, Linear System Theory and Design, Holt, Rinehart and Winston, New York, 1984.

[2] Luenberger D G, An Introduction to Observers, IEEE Trans on Aut Control, Vol AC-16, pp 596-602, 1971.

[3] Owens, D H, Feedback and Multivariable Systems, Peter Peregrinus, Stevenage, England, 1978.

[4] Owens, D H, Multivariable and Optimal Systems, Academic Press, London, 1981.

[5] Owens, D H and Virk, G S, On Sensitivity, Compensation, and the Approach to the Asymptotes in Multivariable Root-Loci, IMA Journal of Math Control and Information, Vol 1, pp 199-221, 1984.

[6] Rosenbrock, H H, Computer Aided Control System Design, Academic Press, Information, Vol 1, pp 199-221, 1974.

[7] Rosenbrock, H H, State-space and Multivariable Theory, Nelson, London, 1970.

Chapter 3

Automatic tuning of commercial PID controllers

P. J. Gawthrop

1. INTRODUCTION.

Most industrial processes are controlled using Proportional-Integral-Derivative (PID) controllers. The popularity of PID controllers can be attributed partly to their robust performance in a wide range of operating conditions, and partly to their functional simplicity which allows process engineers to operate them in a simple and straightforward manner. To implement such a controller three parameters have to be determined for the given process: proportional gain, integral time constant and derivative time constant. Usually, process engineers must tune Proportional-Integral-Derivative (PID) controllers manually, an operation which, if done diligently, can take considerable time. The problem is exacerbated by interactions with other loops.

To overcome this problem several methods have been developed that automatically tune PID controllers. The most well-known method is that of Ziegler and Nichols developed back in 1942-1943, which determines the parameters by observing the gain at which the plant becomes oscillatory, and the frequency of this oscillation. A more useful extension of the method allows the determination of those parameters from the observation of the open-loop response of the plant to a step input change. A similar method has been developed by Hazebroek and Waerden (1950). Several other similar simple methods have been developed since, which automatically generate a special input to the process, and by observing its response, they determine the PID parameters (for example Field 1962, Nishikawa et al 1984, Astrom and Hagglund 1984b). The simplicity of all those methods has made possible the development by the process-controllers industry of a wide range of such 'auto-tuning' instruments which have recently appeared in the market (Morris 1987). Other approaches are given by (Gallier and Otto 1968, Andreiev 1981, Hawk 1983, Proudfoot ,Gawthrop and Jacobs 1983).

Apart from the 'initial tuning' problem, it is also true that the PID algorithm, by being so simple, cannot always effectively control systems with changing parameters, and so may need frequent manual on-line retuning. As such 'difficult' systems are common place in real industrial plants, the need for algorithms which could cope with them encouraged the development of the Adaptive Control methods.

Although the PID algorithm itself is essentially simple, commercial implementations of PID controllers contain a multitude of additional features which embody

By kind permission of the IEEE, this chapter is based on a paper of the same name appearing in the IEEE Control systems magazine, January 1990.

the experience of many years of application. For example integral windup prevention, integral preload, derivative limiting and bumpless transfer are all features essential to safe and effective operation in practice. It is therefore prudent to retain those features when introducing self-tuning; basically sound self-tuning methods have failed because these vital details have not been attended to.

To take account of these points, the approach taken here is to use a standard commercial PID controller (a Eurotherm 820) and use a continuous-time self-tuning PID controller (Gawthrop 1982b, 1986, 1987) to generate its *parameters* as opposed to using the self-tuning PID controller to generate a *control signal.* A similar approach is taken by Koivo and Sorvari (1985). The self-tuning PID controller is implemented on an Atari PC and communicates via an RS-232 port using the Eurotherm communications protocol. This approach has several advantages.

Firstly, the actual control signal is generated by a standard industrial PID controller which, if well-tuned, is a well known and trusted instrument giving guaranteed high performance for a given range of operating conditions. It is when that range changes that the tuner intervenes and re-adjusts the instrument to the new requirements.

Secondly, large processes usually require a large number of smaller control loops for which the performance requirements are such that a standard PID controller is sufficient. In that case it is more reasonable to introduce few self-tuners on a higher supervisory level which will only adjust several controllers each, especially in the more critical loops. Thus the self-tuners will become part of a more complete hierarchical structure. Moreover, in cases where PID controllers have already been installed, it is easier to add a set of supervisory controllers instead of re-building the whole structure.

Thirdly, this approach improves the safety of the plant. If for any reason the adaptive algorithm drifted to unstable conditions, the PID can be dropped back to a set of safe parameters, which will guarantee that the plant will not go unstable. This is evidently much better than dropping back to manual control, which is the commonly used alternative.

Fourthly the start-up of the plant can be greatly improved. Instead of injecting 'artificial' perturbations to the plant to allow the tuner to tune properly, we could simply let the standard PID to drive the plant until the tuner settles to an acceptable set of parameters.

The standard PID algorithm is a single-loop controller. However, most of the real industrial processes are multiloop, with several inputs which affect a number of outputs. Changes in any one of the process loops usually affect several other loops, and in such cases the PID algorithms treat those interaction disturbances as any other disturbance. The need has, therefore, arisen for another algorithm that will be able to tune standard PID controllers and at the same time will provide a means of decoupling any disturbance caused by interacting loops. Such an algorithm will be presented in this paper.

The problem in the case of multiloop processes is to decouple the interaction disturbance. Several methods have been developed for such cases, for example (Borison 1979, Koivo 1980) are two of the most well known. However, in the majority of those algorithms the design is based on a single-controller representation, where the system is represented in matrix transfer function form and the controller is built accordingly. Arrays of process variables are polled at each iteration and multidimensional arrays of parameters are updated to derive the array of the control signals. Such an approach requires excessive computing power, a fact that restricts the efficiency of those controllers.

It can be argued (Morris et al 1982, Nomikos 1988, Gawthrop 1987) that a more effective alternative approach is to split the overall system representation into smaller single-loop units, with the interaction represented as additional transfer-function terms, and to design individual controllers for each loop which will be fed with information from the adjacent loops in order to decouple the interaction disturbance. Such a design will result in several MISO controllers which individually require less computing power than before, and moreover it will be easier to adapt the design according to structural plant modifications.

A practical problem addressed and solved here is to provide for some means of decoupling the interaction, in addition to the tuning of the PID parameters. The difficulty arises from the fact that the control signal is not generated from the tuner, which could identify the interaction and adjust such a signal to decouple it, but is generated by an external instrument which, by itself, cannot take such action. The solution to this problem is to calculate a feedforward signal which will be targeted to cancel the effects of interactions and add this signal to the control signal generated by the PID using the standard 'bias' facility provided in the Eurotherm 820.

2. THE CONTINUOUS-TIME SELF-TUNING PID CONTROL ALGORITHM.

This section provides a brief overview of the continuous-time self-tuning control approach, more details are given in Gawthrop (1986, 1987) and in Nomikos (1988). In the context of adaptive control, the controller design can be performed in either the discrete or the continuous time domain. It is true that most of the contemporary algorithms are based on the former design. However, it has been argued (Edmunds 1976, Astrom et al 1980, Goodwin et al 1986, M'Saad et al 1985) that this approach causes several problems such as the migration into the instability region of the zeros of systems with relative order higher than two, and the fact that important system dynamics which are apparent in a continuous-time representation may be lost in a discrete design. In particular, it is easier to relate the effects of proportional and integral action to the process dynamics if the self-tuning controller design is based on a continuous-time formulation than when it is based on a discrete-time formulation. It has been shown in Gawthrop (1982a, 1987) that it is possible to design the adaptive controller design in a continuous-time context, and then discretise the algorithm for computer implementation. This is in contrast to the discrete-design methods, where the discretisation takes place *before* the controller design.

The continuous-time approach is summarised in the remainder of this section. More details appear in (Gawthrop 1982a, 1986, 1987). The continuous-time design is based on a system model of the form shown where y(s), u(s) and z(s) are the Laplace transforms of the system output, input and noise disturbance respectively and A(s), B(s) and C(s) are polynomials in the s-domain.

$$A(s)y(s) = B(s)u(s) + C(s)z(s) \qquad (1)$$

C(s) is a design polynomial, whereas A(s) and B(s) represent the system dynamics which are either known (non-adaptive design) or unknown (adaptive design).

For processes with high relative order (deg(A) much greater than deg(B)) we may introduce an unrealisable quantity which will emulate the effect of taking pure derivatives as

$$\phi(s) = P(s)y(s) \qquad (2)$$

$$deg(P) = deg(A) - deg(B) \qquad (3)$$

We term this unrealisable quantity $\phi(s)$ the *predictive* term. Since it is unrealisable, it can be shown that we may approximate it by introducing the quantity

$$\phi^*(s) = \frac{F(s)}{C(s)}y(s) + \frac{G(s)}{C(s)}u(s) \qquad (4)$$

The polynomials F(s) and G(s) are given (in the non-adaptive case) from

$$\frac{P(s)C(s)}{A(s)} = E(s) + \frac{F(s)}{A(s)} \tag{5}$$

$$G(s) = E(s)B(s) \tag{6}$$

The polynomial Q(s) is another design polynomial called the *detuning* polynomial. For the moment we assume that Q(s) = 0.

In the adaptive case F(s) and G(s) can be directly estimated using a standard Least-Squares estimator if (4) is expressed in the form shown where $\underline{X}^T$ is the data vector and $\underline{\theta}$ is the vector of the corresponding parameters.

$$\phi^*(s) = \underline{X}^T\underline{\theta} \tag{7}$$

$$\underline{X} = [\underline{X}^y, \underline{X}^u] \tag{8}$$

$$\underline{X}^y = \Big[1,s,...,s^n\Big]y(s)/C(s) \tag{9}$$

$$\underline{X}^u = \Big[1,s,...,s^n\Big]u(s)/C(s) \tag{10}$$

2.1. The Modified System Model

The algorithm presented above has been modified in a way that allows the on-line estimation of PID parameters (Gawthrop 1986). The motive for such a modification was the need to introduce integral action into the algorithm, in order to remove offset errors from the steady-state value of the controlled process. Some researchers have tried to introduce such integral action by forcing an integrator into the controller (for example Clarke et al 1983, and Bellanger 1983). The approach used here is the reverse: the system is modelled to include an offset term, and the corresponding controller is found to have integral action. This approach was also adopted by Scattolini (1986).

The assumption upon which the modification of the algorithm is based mainly concerns the disturbance model. It is argued that real disturbances do not usually have a zero mean; such a disturbance may drift with time. We can model such a process by the following equations where z(s) is a zero-mean disturbance.

$$z'(s) = \Big[(1+cs)/s\Big]z(s) \tag{11}$$

$$A'(s)y(s) = B'(s)u(s) + C'(s)z'(s) \tag{12}$$

For the model (12) to correspond with the original system model (1) the following relations must hold

$$\cdot \quad A(s) = sA'(s) \tag{13}$$

$$B(s) = sB'(s)$$

$$C(s) = (1+cs)C'(s)$$

The integrator in the disturbance model therefore implies that

$$A(0) = B(0) = 0 \tag{14}$$

$$C(0) \neq 0$$

2.2. The Zero-Gain Predictor

The offset problem may be due not only to a non-zero mean disturbance, but also to incorrect choice of the controller parameters. To overcome this, we simply have to constrain the design polynomials to

$$P(0) = 1 \text{ and } Q(0) = 0 \tag{15}$$

This means that be P(s) can be expressed as

$$P(s) = 1 + sP_0(s) \tag{16}$$

$$P_0(s) = p_1 + p_2 s + \cdots + p_n s^{n-1}; \quad n = deg(P) \tag{17}$$

Therefore $\phi(s)$ becomes

$$\phi(s) = y(s) + \phi_0(s) \tag{18}$$

$$\phi_0(s) = sP_0(s)y(s) \tag{19}$$

Eqn. (5) is now modified to

$$\left[(P(s) - 1)C(s)\right]/A(s) = \left[sP_0(s)C(s)\right]/A(s) = E'(s) + F'(s)/A(s) \tag{20}$$

Since A(0)=0, it follows that $F'(0)=0$, and also by defining $G'(s)=B(s)E'(s)$ we will have the following where $F_0(s)$ and $G_0(s)$ have a similar form to $P_0(s)$ of (17).

$$F'(s) = sF_0(s); \quad G'(s) = sG_0(s) \tag{21}$$

In a manner similar to the eqn. (4) it can be shown that the predictor $\phi_0(s)$ is given by

$$\phi_0^*(s) = sF_0(s)y(s)/C(s) + sG_0(s)u(s)/C(s) \tag{22}$$

Since each of the two filters in the predictor has zero steady-state gain, this is termed the "zero-gain predictor".

If we wish to make the prediction $\phi_0^*(s)$ follow the setpoint then a suitable control law will follow from

$$\phi_0^*(s) = w(s) - y(s) \tag{23}$$

to become

$$u(s) = [C(s)/G_0(s)] \left[[w(s)-y(s)]/s - [F_0(s)/C(s)]y(s)\right] \tag{24}$$

This controller has the desirable integral action, and will be used as the basis of the PI and PID controllers to be examined.

In Gawthrop (1986) it is also proved that this zero-gain predictor can be modelled in a state-space representation very similar to that explained in section (2.1), and is therefore suitable for the least-squares estimator.

Although this form has integral action, it is not yet readily transformable to a PID-like control law. Another alternative would be to choose C(s) to be such that C(0)=1, that is

$$C(s) = 1 + sC_0(s) \tag{25}$$

Adding the term $sC_0(s)/C(s)y(s)$ to each side of (22) we find that

$$\phi_0^{c*}(s) = sF_0^c(s)y(s)/C(s) + sG_0(s)u(s)/C(s) \tag{26}$$

It is shown in Gawthrop (1986) that the quantity $\phi_0^{c*}(s)$ can be regarded as the prediction of

$$\phi^c(s) = w(s) - y(s)/C(s) \tag{27}$$

Substituting for this alternative zero gain predictor in a relation similar to (23), we will find a control law

$$u(s) = G_0(s)^{-1} \left[[w(s){-}y(s)]/s + C_0(s)w(s) - F_0^c(s)y(s) \right] \tag{28}$$

The polynomials $F_0^c(s)$ and $G_0^c(s)$ are given by expressions very similar to (20), and again it can be shown that this predictor as well can be expressed in a form suitable for estimation.

2.3. PI and PID Controllers

In the previous section we have seen how we can introduce the desirable integral action in the control law (28). However, this law is not yet similar to a standard PID controller. To achieve this we need to restrict the system polynomials. In particular, the assumptions (12) coupled with the conditions

$$deg(A') = 1 \; and \quad deg(B') = 0 \tag{29}$$

will be shown to give a PI controller, and coupled with

$$deg(A') = 2 \; and \quad deg(B') = 0 \tag{30}$$

will be shown to give a PID controller. It is interesting to notice that this structure is also in accordance with the "second-order system with delay" structure used by most of the researchers referenced in the introduction for their adaptive PID algorithms.

It should be pointed out that assumptions (29-30) do not restrict the class of the real systems (to which this algorithm can be applied) to be of second order. They only restrict the 'nominal' system model. The real plant might be (and usually is) of higher order and the difference can be modelled (or approximated) in the some 'neglected dynamics' terms. The importance of those terms can be investigated using the techniques presented in Gawthrop (1987) where a method of investigating the robustness of this approach in the face of neglected high-frequency dynamics is analysed. If those terms are found to destabilize the performance of the algorithm then appropriate corrective action should be taken, by modifying the controller design polynomials.

The assumptions (29) combined with the above described predictor structure will give the following where b, c, e and f^c are real numbers.

$$C(s) = 1{+}cs \tag{31}$$

$$E'(s) = e$$

$$F_0^c(s) = f^c$$

$$G_0(s) = eb$$

The control law (28) will then become

$$u(s) = (eb)^{-1} \left[[w(s){-}y(s)]/s + cw(s) - f^c(s)y(s) \right] \tag{32}$$

which is similar to a standard PI law where k is the proportional gain, T_i the integral time constant and kr_p the proportional gain on the setpoint.

$$u(s) = k \left[[w(s){-}y(s)]/(T_i s) + r_p w(s){-}y(s) \right] \tag{33}$$

This similarity is illustrated by the relations

$$T_i = f^c \tag{34}$$

$$k = f^c/eb$$

$$r_p = c/f^c$$

If now we assume the system structure of (30), then the controller polynomials will be of the form

$$C(s) = c_0 s^2 + c_1 s + 1$$
$$E'(s) = e_0 s + e_1 \tag{35}$$
$$F_0{}^c(s) = f_0{}^c s + f_1{}^c$$
$$G_0(s) = be_0 s + be_1$$

The control law will then become

$$u(s) = \left[(e_1 + e_0 s)b\right]^{-1}\left[[w(s)-y(s)]/s + c_1 w(s) - f^c{}_1 y(s) + (c_0 w(s) - f^c{}_0 y(s))s\right] \tag{36}$$

A standard PID law is typically

$$u(s) = k\left[[w(s)-y(s)]/T_i s + (r_p w(s)-y(s)) + [r_d w(s)-y(s)]T_d s/(1+T_f s)\right] \tag{37}$$

where in addition to (33), T_f is now the time constant of a filter to avoid derivative action at high frequencies, T_d is the derivative time constant, and r_d controls the derivative 'kick'. Comparing those two control laws we see that they are equivalent if the following relations apply

$$T_f = be_1/be_0 \tag{38}$$
$$T_i = f_1{}^c - T_f$$
$$T_d = f_0{}^c - (T_i T_f)/T_i$$
$$k = T_i/be_1$$

Relations (38) are used in the algorithm to derive on-line PID parameters corresponding to the estimated tuner parameters, and tune external PID controllers.

It is important to note that equation 37 does not represent the only possible PID controller structure. For example, an alternative form is given by the following (where the proportional and derivative 'kick' terms have been ommited for simplicity).

$$u(s) = k\left[1 + 1/T_i s\right]\left[T_d s/(1+T_f s)\right][w(s) - y(s)] \tag{37a}$$

Equations such as 37a would lead to a new set of conversion equations to replace equations 38.

2.4. The multivariable version of the algorithm.

As explained in the introduction, in order to apply the single-loop algorithm that we presented above to multiloop processes we need to introduce a way to decouple the interaction disturbance. As shown in (Gawthrop 1985), a possible representation of one loop of such a multivariable process is described by Figure 1. The system and interaction transfer function are described as

$$S_{ij}(s) = B_{ij}(s)/A_{ij}(s); \quad i=1,...,n; \quad j=1,...,n \tag{39}$$

To satisfy conditions that will be presented later the assumption must be made that the interaction transfer functions are constants, ie.

$$A_{ij}(s) = 1 \tag{40}$$
$$S_{ij}(s) = B_{ij}(s) = const; \quad i \neq j$$

Combining (39) with (11-13) we will find that the extended system representation is

$$A'_{ii}(s)y_i(s) = B'_{ii}(s)(u_i(s) - \sum_{i \neq j}^{n} S_{ij}(s)y_j(s)) + C'_i(s)z'_i(s) \tag{41}$$

Considering (13) and (21) we can define for each such loop

$$G_j'(s) = S_{ij}(s)E_i(s)B_{ii}(s) = sS_{ij}(s)G_{i0}(s) = sG_{j0}(s) \tag{42}$$

The extended form of the zero-gain predictor (26) will be

$$\phi_{i0}^{c*}(s) = sF_{i0}^c(s)/C_i(s)y_i(s) + sG_{i0}(s)/C_i(s)u_i(s) - \sum_{i \neq j}^{n} sG_{j0}(s)/C_i(s)y_j(s) \tag{43}$$

leading to the extension of the control law (28) as

$$u_i(s) = 1/G_{i0}(s)\left[[w_i(s)-y_i(s)]/s + C_{i0}(s)w(s) - F_{i0}^c(s)y_i(s) \right] + \sum_{i \neq j}^{n} \frac{G_{j0}(s)}{G_{i0}(s)}y_j(s) \tag{44}$$

It is interesting to notice from (42) that, for each interaction term, the ratio following is the same as the interaction transfer function itself.

$$G_{j0}(s)/G_{i0}(s)y_j(s) = S_{ij}(s) \tag{45}$$

This means that the sum of those ratios in (44) corresponds physically to the opposite of the disturbance from the interacting loops. Taking into account also the assumption (30) we need to have assumed (40) so that those transfer functions are realisable.

The control signal of (44) has therefore to compensate for two factors: The first is the dynamics of the loop under control (it must reduce the output-setpoint error to zero), and the second is the disturbance from the interaction.

Taking, therefore, assumptions (30), for the PID-controller case the final extended control law will become

$$u_i(s) = \left[e_{i1} + e_{i0}s)b_i \right]^{-1} \left[[w_i(s) - y_i(s)]/sc_{i1}w(s) - f^c_{i1}y_i(s) + (c_{i0}w_i(s) - f^c_{i0}y_i(s))s \right]y_i(s) \tag{46}$$

$$+ \sum_{i \neq j}^{n} \frac{g_{j0}s+g_{j1}}{g_{i0}s+g_{i1}}y_j(s)$$

This control law is consisted of two parts. The first part is exactly the same as (28). Based on this part the PID parameters are derived in exactly the same way as in section 1.4. Then a feedforward Bias signal is calculated based on the second part of (46), which would compensate for the interaction disturbance. It is given by

$$b_i(s) = \sum_{i \neq j}^{n} g_{j0}s+g_{j1}/g_{i0}s+g_{i1}y_j(s) \tag{47}$$

If this signal is sent as a feedforward signal to the external PID controllers, the result will be the desired decoupling of the interaction. In the case of the PI controllers (assumptions 29) the design is very similar to the PID case.

There is also scope to introduce a non-zero detuning polynomial Q(s) (from eqn. 5). It has been shown in (Gawthrop 1987) that such an action will enhance the stability of the algorithm. A possible choice for Q(s) can be

$$Q(s) = [q_0s^2 + q_1s]/C(s) \tag{48}$$

leading to a detuned version of the control law (46) as

$$u_i(s) = \left[g_{i1} + q_{i1}+(g_{i0} + q_{i0})s \right]^{-1} \left[w_i(s)-y_i(s)]/sc_{i1}w(s) - f^c_{i1}y_i(s) + (c_{i0}w_i(s) - f^c_{i0}y_i(s))s \right] \tag{49}$$

$$+ \sum_{i \neq j}^{n} \frac{g_{j0}s+g_{j1}}{(g_{i0} + q_{i0})s+g_{i1} + q_{i1}}y_j(s)$$

So now the detuned feedforward bias factor becomes

$$b_i(s) = \sum_{i \neq j}^{n} \frac{g_{j0}s + g_{j1}}{(g_{i0} + q_{i0})s + g_{i1} + q_{i1}} y_j(s)$$ (50)

whereas the estimated PID gain k and derivative filter time constant T_f from (38) become

$$T_f = be_1 + q_1/be_0 + q_0; \quad T_i = f_1{}^c - T_f; \quad k = T_i/be_1 + q_1$$ (51)

In the non-adaptive case (where the system polynomials $B_{ij}(s)$ and $A_{ij}(s)$ are assumed known) the parameters of the polynomials $F_{i0}{}^c(s)$, $G_{i0}(s)$ and $G_{j0}(s)$ can be calculated from relations similar to (5) and then we can solve for the PID parameters from (38) or (51). In the adaptive case (unknown system polynomials) eqn.(43) can be expressed in the standard form

$$\phi_{i0}{}^{c*}(s) = \underline{X}^T \underline{\theta}$$ (52)

where now the data vector $\underline{X}^T$ is composed of

$$\underline{X}^T = [\underline{X}_u; \underline{X}_{y_1(s)}; \dots; \underline{X}_{yn}]$$

$$\underline{X}_u = C_i(s)^{-1} [u_i(s), su_i(s), \dots, s^n u_i(s)]$$ (53)

$$\underline{X}_{yj} = C_i(s)^{-1} [y_j(s), sy_j(s), \dots, s^n y_j(s)], \quad j=1..n$$

and the parameter vector $\underline{\theta}$ is composed of the vectors of the parameters corresponding to each element of $\underline{X}^T$. The i factors of those vectors correspond to the main loop dynamics and the j factors correspond to the dynamics of the interaction from the other loops.

$$\underline{\theta}^T = [\underline{\theta}_u; \underline{\theta}_{y_1(s)}; \dots; \underline{\theta}_{yn}]$$ (54)

A standard Least-Squares estimation algorithm is used to estimate $\hat{\underline{\theta}}$ and the estimated parameters are used in (38) or (51) to give the estimated PID coefficients.

3. FIRST APPLICATION: A SET OF LABORATORY TANKS.

Two sets of experiments on real processes are presented in the paper. This section describes the first: a set of mutually coupled tanks where the water level was under control. The second process (an industrial plastics extruder barrel) is discussed the next section. Further details of both these applications can be found in a doctoral thesis (Nomikos 1988).

3.1. Experimental Equipment

3.1.1. The tanks

A set of four adjacent tanks were constructed out of perspex and fixed above a sump of the same material. The tanks have no internal connections, but may be interconnected in a variety of configurations by means of rubber tubing, clamps and valves. The tanks are consecutively numbered from 1 to 4 starting at the left. For the single-loop experiment, tanks 1 and 2 were connected and tank 2 discharges into the sump; for the two-loop experiment, tanks 1 and 2, 2 and 3, and 3 and 4 were connected. Tanks 2 and 3 also discharged into the sump. This discharge introduced a nonlinearity into the system due to the non-linear pressure-flow characteristic.

3.1.2. The transducers

Each tank was equipped with a silicon pressure sensor tightly sealed into the upper end of a plastic tube whose lower end reached the bottom of the tank. Thus level was indirectly measured from the pressure at the bottom of each tank. The pressure sensor output was suitably conditioned and buffered by an instrumentation amplifier to give a 0-10v signal. For the single-loop experiment, the tank 2 level was used as output. For the two loop experiment tank 2 and tank 3 levels were used as output. It was experimentally verified that the level to measured voltage signal was linear.

The 12vdc submersible bilge pumps were placed in the sump and could discharge, via rubber tubes, into any of the four tanks. They were each driven, via buffers and power amplifiers, by a 0-10v signal. For the single-loop experiments, pump 1 discharged into tank 1 and pump 2 was not used. For the two-loop experiment pump 1 discharged into tank 1 and pump 2 into tank 4. In both cases, the discharge end of the tube was above the water surface. Experiments showed that the pumps showed a highly non-linear, and dynamic, characteristic relating input voltage to flow rate. At low voltages the pump did not deliver enough pressure to drive any water through the tube; at high voltages the flow rate became independent of voltage. When switching from low to high voltage, there was a significant delay before the tube filled with water.

3.1.3. The Controllers

The PID controllers were industry standard Eurotherm 820 controllers equipped with a 0-10v interface and adjusted so that 0-10v corresponded to a 0-100% range as both input and output. These controller operated with a 50ms sample interval and had state of the art integral desaturation, bumpless transfer and user interface. The particular feature of interest here was the ability to send P, I and D values, and receive measured variables and control signals via a standard RS232 link and an appropriate message protocol. For the two loop experiments, the additional feature of adding an additional bias signal to the 820 control signal was utilised.

For the single-loop experiment, controller 1 measured the level in tank 2 and sent the control signal to pump 1; controller 2 was not used. For the two-loop experiment, controller 1 was connected as for the single-loop experiment while controller 2 measured the level in tank 3 and sent the control signal to pump 2.

3.1.4. The Tuner

The self-tuning algorithm of section 2 was implemented in Pascal on an Atari 1040 ST personal computer. The PID parameters were obtained from equations 5 1 and the proportional gain converted into proportional band (=100/proportional gain).

The communications protocol was implemented in C and linked to the Pascal program. The signals were requested from the 820 controller parameters updated and PID values returned with a two second cycle time. It is important to note that the control signal itself was generated at the 820 sample interval of 50 ms. Data was displayed on the screen and also written to RAM disc for later copying onto the floppy discs at the end of each experiment.

3.2. Mathematical Model

An idealised mathematical model of the apparatus is presented in Figure 2, where A is the area of the water surface, h_i is the level in each tank, k_p is the constant associated with the pump, k_h are the constants of the output holes and the pipes, k_s is the pressure sensor constant. The signal y' is the output of the other interacting loop. There are two key features to note. Firstly, the relationship between inflow into tank 1 (or 4) and level in tank 2 (or 3) is predominantly second order. Secondly, the interaction is from output of tank 3 to tank 2, not to tank 1. Thus the interactive model used (output - input) involves the inverse of the transfer function relating the inflow to tank 1 to the inflow to tank 2 from tank 1.

The implication of this improper interaction model is examined in the sequel.

3.3. Single loop experiments.

3.3.1. Purpose

The main purpose of these experiments was to verify our claim that the use of the self-tuning controller allows the user to specify closest-loop performance in terms of desired response $1/P(s)$. The PID is automatically tuned to give this desired response.

3.3.2. Description

In each case, we chose a second-order dead-beat response specified by

$$1/P(s) = 1/(1 + sT) \tag{55}$$

A series of 5 experiments was conducted for $T=10$, 15, 20, 30 and 50 sec. Figure 3 shows the results for $T=20$*. The Figure has two sets of graphs. The upper graph shows three signals: the firm square wave signal is the controller set point, the dashed line is that setpoint passed through the desired system $1/P(s)$ and the other firm line is the measured output signal. All three quantities are expressed as percentage of full scale. The lower graph shows the control signal expressed as percentage of full scale. The results for other other values of T have a similar form but the both the desired and actual output time constants decrease, and the corresponding control signal increases, with T. Thus T acts as a performance-orientated parameter trading off speed of response against control signal magnitude whilst retaining the same form of response.

Figure 4 shows the proportional band for each of the 5 experiments: the largest value corresponds to $T=50$ the smallest to $T=10$. A expected, the proportional band decreases with desired time-constant T. The corresponding integral time constants have similar behaviour except that they increase with T; T_i converges to about 3.8sec when $T=10$ and to about 3.1sec when $T=50$. The derivative time constant T_d converges to about 1.1sec for each value of T.

3.3.3. Discussion

The overall trend is clear: the PID parameters are adjusted so that the closed- loop system is approximately as desired. The responses differ in detail from their desired values, particularly for large and small T. The former is explained by the low controller gains (large proportional band) not being sufficient to overcome non- linear behaviour; the latter is explained by control signal saturation.

Despite the non-ideal system under control, the behaviour is broadly as

---

* Figures corresponding to the other experiments are given in an internal report (Gawthrop & Nomikos, 1989).

advertised.

3.4. Two-loop experiments

3.4.1. Purpose

The main purpose of this sequence of experiments was to investigate the advantages of explicitly identifying process interaction as a means of reducing closed-loop interaction. In addition, we wished to verify our approach of using essentially single- loop instruments in a multi-loop setting by computing an additional 'bias' signal.

In hindsight, the presented sequence of experiments also illustrate the power of the continuous-time approach in analysing and explaining our initial failure and devising methods to overcome the problem. In particular, the result noted by Gawthrop (1987 and 1985) that the poor robustness of self-tuning controllers to neglected dynamics can be overcome by weighting the control signal, is used to good effect.

3.4.2. Description

As in the single-loop case, each loop had a desired response specified by

$$1/P(s) = 1/(1 + sT) \tag{56}$$

As we wished to investigate the effect of interaction, we chose the 'worst case' of the time constants used in the single-loop case: $T=10sec$. A series of experiments was performed as follows:

1   Untuned PID - no interaction term. The PID coefficients used are those generated in the corresponding single-loop experiment.

2   Tuned PID - no interaction term

3   Tuned PID - standard interaction term

4   Tuned PID - standard interaction term with weighting (Q = 0.1)

5   Tuned PID - setpoint interaction term

In each case, loop one is driven by a square setpoint with period 600sec and loop two is driven by a step change. Thus the effects of interaction are most clearly seen on loop 2.

Figure 5 is of the same form as Figure 3 except that the lower graph contains one extra signal: a dashed line corresponding to the computed interaction, or 'bias' term. It shows the signals corresponding to loop 2 for the 5th experiment*. The results of the other experiments can be described relative to this figure as follows, in each case the description refers to loop 2 whose output would ideally settle to a constant value of 12.

1   A deviation of ±2 units appears on output 2 when output 1 changes. The control signal is a noisy square wave moving between 60 and 100 in unison with output 1. The computed interaction is zero.

2   The results are similar in form to those of experiment 1 except that the deviation gradually reduces to about ±0.75 and the control signal is rather noisier.

3   The control signal exhibits large oscillations at time 500 and the experiment was terminated.

---

* Figures corresponding to the other experiments are given in an internal report (Gawthrop & Nomikos, 1989).

4   The deviations in the output are of similar magnitude to those in experiment 2, and the computed interaction term moves in unison with output 1.

5   As depicted in Figure 5, the output deviations are reduced to about +−0.5, and the computed interaction term is not noisy. The phase-advance nature of this interaction transfer function is clear.

3.4.3. Discussion

Comparison of experiments 1 and 2 indicates that the tuner itself is able to reduce the effect of interaction due to an adjacent loop and this is itself an encouraging result.

The result of experiment 3 indicates a pitfall of this approach: the neglected dynamics implicit in the modelling procedure lead to the tuning procedure attempting to approximate an improper interaction term with a high-gain proper term - with the unfortunate result observed. However, as discussed in Gawthrop (1987) and Gawthrop (1985) self-tuning controllers are sensitive to neglected dynamics unless control weighting is used: this motivates experiment 4. Note that the 'bias' signal is now stable, but quite noisy.

As discussed in Nomikos (1988) the noise amplification inherent in the generation of the bias term can be avoided by replacing the output by a clean approximation, in this case the setpoint itself. The result is shown in Figure 5, where the bias signal is smoother and yet the interaction effect is reduced.

Thus the single-loop approach to satisfactory in the two-loop situation; but well designed decoupling gives improved performance. Because all the terms within the control and tuning algorithms are in a continuous time format they have a clear interpretation to the user and thus, as illustrated in this application, the approach lends itself to algorithm engineering.

4. SECOND APPLICATION: TEMPERATURE CONTROL OF A PLASTICS EXTRUDER

4.1. Problems in controlling extrusion processes.

Plastic extrusion is a well established method widely used in the polymer processing industry (Amrehn 1977). Such extruders typically consist of a large barrel divided in several temperature zones, with a hopper at the one end, and a die at the other. Polymer is fed into the barrel in raw and solid form from the hopper, and is pushed forward by a powerful screw. Simultaneously it is gradually heated while passing through the various temperature zones set in gradually increasing temperatures. The heat produced by the heaters in the barrel, together with the heat released from the friction between the raw polymer and the surfaces of the barrel and the screw, eventually cause the melting of the polymer, which is then pushed by the screw out from the die, to be further processed for various purposes.

Such a process is highly non-linear, and a typical example of a difficult control problem (Kochhar and Parnaby 1977). The output variables typically are the outflow from the die, or equivalently the pressure in the vicinity of the die, and the polymer temperature. The main controlling variable is the screw speed, since the response of the process to it is instantaneous. The input signal to the heaters is a secondary variable, mainly because the thermal inertia of the barrel seriously dampens the response of the barrel to such changes. Up to date, only approximate modelling techniques have been proposed, and a complete overall modelling, designing and controlling strategy has yet to appear (Kochhar and Parnaby 1978, Costin et al 1982).

Control is still exercised mainly by typical PID controllers, which usually tolerate the system for short ranges of operating conditions, but which need frequent retuning as the operating and environmental conditions change with time. Thus those controllers must be constantly supervised.

This paper does not attempt to give a total solution to the overall control problem of the extrusion process. Only the temperature control of the barrel is considered here, because the screw speed was available for manual control only. The aim of the experiment is to prove that the tuner estimates sensible parameters for a PID controller in a real industrial situation, although the final result is very much prejudiced by the ineffectiveness of temperature control. .

The specific process which we are dealing with is a typical extruder barrel (See Figure 6) with four main temperature zones and two more close to the die, used to extract plastic to be formed into various fibres and cables. The plant was installed in an industrial site.

The temperature of the barrel had been controlled by six standard purpose-built industrial PID controllers (Eurotherm 820), one for each zone, operating on the same standard PID algorithm as for the coupled tanks. Several limits and alarm settings were the only difference from the previous experiments. The Self-Tuner was again running from the same Atari ST, in exactly the same way as described in previous sections. Some alternative algorithms are presented in (Gawthrop et al 1988).

4.2. Zero screw speed.

4.2.1. Purpose

When there is no flow through the extruder barrel, the temperature control is essentially that of an inert thermal mass. However, because the system is physically described by a partial differential equation, an essentially *transcendental* transfer function is being controlled on the assumption that the transfer function is *rational* and second order. It is therefore essential that the proposed algorithm is robust in the presence of such approximations. The purpose of this experiment is to verify robustness in this practical example.

4.2.2. Description

Figure 7 is of the same format as Figure 3*, and shows the results using a tuned controller. Note that the overshoot is substantially reduced compared to a similar run with the untuned controller. The estimated Proportional Band settles down after about 100 points; the other PID parameters take about 10 times as long to settle.

4.2.3. Discussion

Despite the approximations inherent in the tuner model, the tuning algorithm is able to give a better control than the untuned controller. The imperfections in the performance appear to be a result of these approximations. We believe that a more complex structure than PID is needed to overcome these defects.

4.3. Variable Screw Speed

---

* Figures corresponding to the other experiments are given in an internal report (Gawthrop & Nomikos, 1989).

4.3.1. Purpose

The rotating screw imparts a large heat flow into the system due to viscous dissipation in the polymer; a variable screw speed therefore imparts a significant disturbance. The purpose of this experiment was to convince the industrial user that the tuner could cope with this effect.

4.3.2. Description

Figure 8 follows the same format as Figure 7*. Note the deviations in temperature are due to changes in screw speed, and that an initial setpoint change is used to aid tuning.

4.3.3. Discussion

Clearly, the self-tuning PID algorithm cannot do better than a well-tuned PID controller, and the results indicate this. However, it has been demonstrated that the tuning algorithm can still operate in the adverse circumstances.

5. CONCLUSIONS

The continuous time self-tuning controller, hitherto used as a complete control algorithm generating a control signal to be sent to the process, has been shown to be also effective in a new role: generating the parameters of a PID controller external to the self-tuning algorithm. These capabilities are illustrated and confirmed by application to both a laboratory scale process and a real industrial plastic extruder.

An important advantage of using the self-tuning algorithm in this mode is that the wealth of experience built in to commercial PID controllers is retained. In this case, the fact that we were using a tried and trusted industrial PID controller, well known to the company involved, was instrumental in obtaining permission to carry out the industrial experiments.

From our experience in the industrial application, we can vouch for the convenience and simplicity of having a personnel computer based automatic tuning facility interfaced to the process via a standard RS232 connection. This could well be the basis of a commercial automatic tuning instrument.

Self-tuning algorithms can be developed either within a discrete time or a continuous time context. The power of the latter approach has manifested itself in this research in two ways. Firstly the algorithm directly generates standard PID parameters, such as proportional band, derivative and integral time constants, directly. Secondly, because all the terms within the control and tuning algorithms are in a continuous time format they have a clear interpretation to the user and thus the algorithm lends itself to algorithm engineering.

As has been pointed out by Astrom (1980), self-tuning algorithms may be distinguished from conventional algorithms by the fact that the user specifies performance, not how to achieve that performance. The particular algorithm given here specifies performance in terms of close loop performance. In particular, the performance of the second order systems considered here was essentially specified by setting the damping ratio equal to one and the time constance of the response was varied. It was demonstrated in the experiment that the algorithm behaved broadly as advertised; the self-tuning algorithm adjusted the PID controller so that the closed loop response had the requested time constant.

As well as its role of automatically tuning feedback terms, the self-tuning algorithm was also used to tune feedforward terms. Because of the constraints the controller actually used we were not able to send feedforward parameters directly to the control algorithm but rather generated the additional control signal directly. However, with suitable hardware we could have passed parameters instead of control signals and this would have had the important advantage that the control system could generate these feedforward signals with the tuning device removed. This is an area for further research and development.

6. ACKNOWLEDGEMENTS

Mr L. Smith of Eurotherm Ltd. provided the Eurotherm 820 controllers, suggested the application to extruder control and gave us advice and encouragement. This work formed part of a project sponsored by SERC under grant GR/E 61854 (ex GR/D 45642). We would also like to thank the editor, associate editor and referee for helpful suggestions for improving the manuscript.

7. REFERENCES

Amrehn H., 1977, "Computer Control in the Polymerization Industry", Automatica, Vol. 13, pp. 533-545.

Andreiev N., 1981, "A New Dimension: A Self-Tuning Controller that Continually Optimizes PID Controllers", Control Engineering, August 1981, pp. 84-85.

Astrom K.J., Hagander P., Sternby J., 1980, "Zeros of Sampled Systems", Automatica, Vol. 20, pp. 31-38.

Wittenmark, B. and Astrom, K.J.: "Simple self-tuning controllers" in Methods and applications in adaptive control ed. H. Unbehauen. Springer 1980.

Astrom K.J., Hagglund T., 1984, "Automatic Tuning of Simple Regulators with Specifications on Phase and Amplitude Margins", Automatica, Vol. 20, No. 5, pp. 645-651.

Bellanger P.R., 1983, "On Type 1 Systems and the Clarke-Gawthrop Regulator", Automatica, Vol. 19, No. 1, pp. 91-94.

Borison U., 1979, "Self-Tuning Regulators for a Class of Multivariable Systems", Automatica, Vol. 15, pp. 209-215.

Clarke D.W., Hodgson A.J.F., Tuffs P.S., 1983, "Offset problem and k-incremental Predictors in Self-Tuning Control", IEE Proceedings, Vol. 130, Part D, No. 5, pp. 217-225.

Costin M.H., Taylor P.A., Wright J.D., 1982, "A Critical Review of Dynamic Modelling and Control of Plasticating Extruders", Polymer Engineering and Science, Vol. 22, No. 7, pp. 393-401.

Edmunds J.M., 1976, "Digital Adaptive Pole-Shifting Regulators", Ph.D. Thesis, UMIST.

Field W.B., 1962, "Adaptive Three-Mode Controller", ISA Journal, Vol. 9, No. 2, pp. 30-33.

Gallier P.W., Otto R.E., 1968, "Self-Tuning Computer Adapts DDC Algorithms", Instrumentation Technology, February 1968, pp. 65-70.

Gawthrop P.J., 1982(a), "A Continuous-Time Approach to Discrete-Time Self-Tuning Control", Optimal Control Applications & Methods, Vol. 3, pp. 399-414.

Gawthrop P.J., 1982(b), "Using the Self-Tuning Controller to tune PID regulators", Sussex University Internal Report CE/T/2.

Gawthrop P.J., 1985, "Robust Self-Tuning Control of n-input n-output systems", 7th IFAC Symposium on Identification and System Parameter Estimation, York, 1985.

Gawthrop P.J., 1986, "Self-Tuning PID Controllers : Algorithms and Implementation", IEEE Transactions on Automatic Control, Vol. AC-31, No. 3.

Gawthrop P.J., 1987, Continuous-Time Self-Tuning Control Volume 1 : Design, Research Studies Press.

Gawthrop P.J., Nomikos P.E., Smith L., 1988, "Adaptive Temperature Control of Industrial Processes - A Comparative Study", IEE Conference Control 88, Oxford, April 1988.

Gawthrop P.J. and Nomikos P.E. 1989, "Automatic Tuning of Commercial PID controllers. Single and multi-loop applications", Control Engineering Report 89.1, Department of Mechanical Engineering, Glasgow University.

Goodwin G.C., Lozanno-Leal R., Mayne D.Q., Middleton R.H., 1986, "Rapprochement between Continuous and Discrete Model Reference Adaptive Control", Automatica, Vol. 22, No. 2, pp. 199-207.

Hawk W.M., 1983, "A Self-Tuning, Self-Contained PID Controller", Proceedings of American Control Conference, San Francisco 1983, pp. 838-842.

Hazebroek P., Van Der Waerden B.L., 1950, "Theoretical Considerations on the Optimum Adjustment of Regulators", ASME Transactions, April 1950, pp. 309-315.

Kochhar A.K., Parnaby J., 1977, "Dynamical Modelling and Control of Plastics Extrusion Processes", Automatica, Vol. 13, pp. 177-183.

Kochhar A.K., Parnaby J., 1978, "Comparison of Stochastic Identification Techniques for Dynamic Modelling of Plastics Extrusion Processes", Proceedings Institute of Mechanical Engineers, Vol. 192, pp. 299-309.

Koivo H.N., 1980, "A Multivariable Self-Tuning Controller", Automatica, Vol. 16, pp. 351-366.

Koivo H.N., Sorvari J., 1985, "On-Line Tuning of a Multivariable PID Controller for Robot Manipulators", Proceedings 24th IEEE Conference on Decision and Control, Fort Lauderdale, USA, pp. 1502-1504.

Morris A.J., Nazer Y., Wood R.K., 1982, "Multivariable Self-Tuning Process Control", Optimal Control Applications and Methods, Vol. 3, pp. 363-387.

Morris H.M., 1987, "How Adaptive are Adaptive Process Controllers?", Control Engineering, March 1987, pp. 96-100.

M'Saad M., Ortega R., Landau J.D., 1985, "Adaptive Controllers for Discrete-time Systems with Arbitrary Zeros: An Overview", Automatica, Vol. 21, No. 4, pp. 413-423.

Nishikawa Y., Sannomiya N., Ohta T., Tanaka H., 1984, "A Method for Auto-Tuning of PID Control Parameters", Automatica, Vol. 20, No. 3, pp. 321-332.

Nomikos P.E., 1988, "Multivariable Self-Tuning Controllers for Industrial Applications", D. Phil. Thesis, University of Sussex.

Proudfoot C.G., Gawthrop P.J., Jacobs O.L.R., 1983, "Self-Tuning PI Control of a pH Neutralisation Process", IEE Proceedings, Vol. 130, Part D, No. 5, pp. 267-272.

Scattolini R., 1986, "A Multivariable Self-Tuning Controller with Integral Action", Automatica, Vol. 22, No. 6, pp. 619-627. Vol. 107, No. 4, pp. 278-283.

Ziegler J.G., Nichols N.B., 1942, "Optimum Settings for Automatic Controllers", ASME Transactions, November 1942, pp. 759-768.

Ziegler J.G., Nichols N.B., 1943, "Process Lags in Automatic Control Circuits", ASME Transactions, July 1943, pp. 433-444.

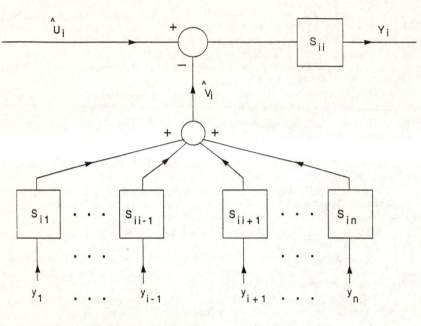

FIGURE 1 The ith subsystem model

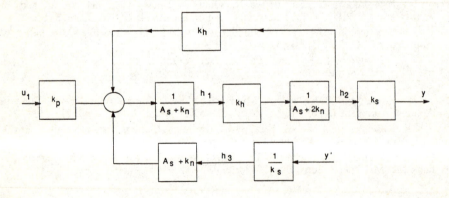

FIGURE 2 The tank system block diagram

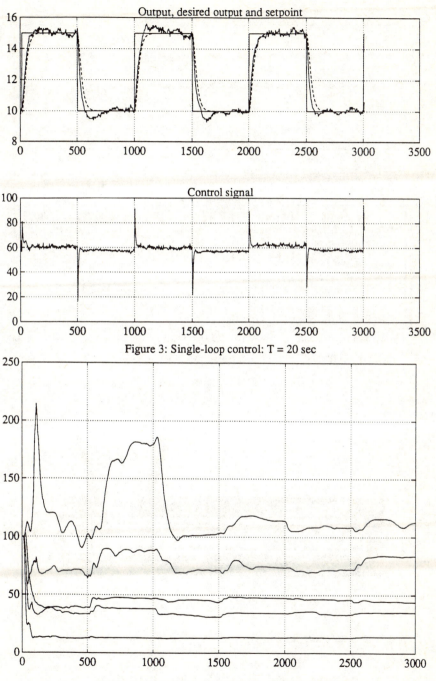

Figure 3: Single-loop control: T = 20 sec

Figure 4: Single-loop control: Proportional band (%)

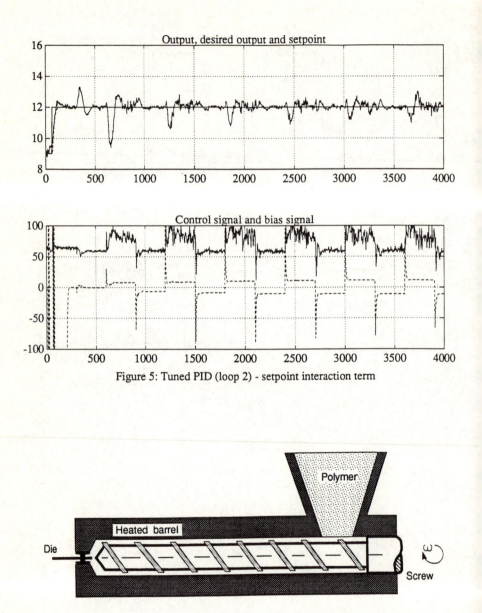

Figure 5: Tuned PID (loop 2) - setpoint interaction term

FIGURE   6 Extruder Barrel

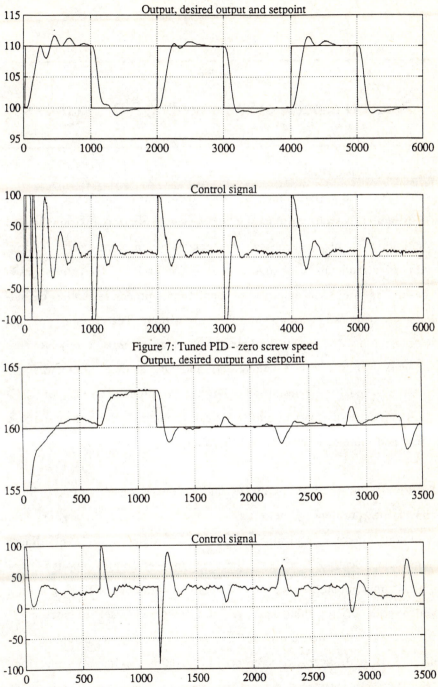

Figure 7: Tuned PID - zero screw speed

Figure 8: Tuned PID - nonzero screw speed

Chapter 4

Practical aspects of implementing PID controllers

L. S. Smith

1. INTRODUCTION

This paper details some of the techniques employed in the implementation and application of process control instrumentation for industrial plant. Process engineers have become very familiar with regulators using proportional + integral + derivative (PID) control, as the vast majority of continuous processes under closed-loop control use variants of this algorithm. The reason for the widespread use of the PID algorithm is that it forms a reliable core for very robust control regulators, and the tuning parameters are relatively easily understood. Digital implementations of PID algorithms are discussed, and practical problems with the associated input and output interfaces are highlighted.

Section 2 covers both hardware and software aspects of the input conversion problem, generating a relatively clean, linear, numeric representation of a typically noisy sampled signal from a non-linear transducer. Section 3 details some of the techniques applied to a digital implementation of the standard continuous PID algorithm, which enhance its robustness and transient response. In section 4, techniques for interfacing to a variety of output actuators are discussed, with specific reference to problems with non-linearities such as those presented by valves. Section 5 gives an overview of the higher level implementation problems, which need to be solved by auto-tuning and adaptive algorithms. Finally, section 6 briefly highlights the impact of communication links on single-loop controllers, and ultimately on overall system design.

2.   INPUT CONVERSION

Sampled data instrumentation encounters the same difficulties processing signals from transducers as met by its analogue predecessors, and adds two further sources of error. These error sources are due to the finite resolution (in amplitude and time) provided by the digital data system, giving scope for lost information.

2.1. Time resolution

The sampling frequency required for digital instrumentation to obtain an error free representation of the continuous signal is given by Shannon's sampling theorem [3], i.e greater than twice the maximum frequency of the sampled signal. In order to avoid high frequency noise on the transducer signal being aliassed into the sampling pass band, the input hardware must include low pass filtering, with a cutoff frequency up to an order of magnitude lower than the sampling frequency. The undesirable effect of aliassing can be illustrated by considering a process signal being sampled at 10 Hz. If this signal includes a noise component at, say, 11 Hz together with the signal of interest from a control viewpoint, the 11 Hz noise will be aliassed down to 1 Hz where it is indistinguishable from the input to be controlled.

From this example it becomes obvious that it is important to know the maximum frequency content of the signal to be controlled, and of the associated noise. For this reason, general purpose process controllers tend to be grouped into application domains such as temperature (<10 Hz sample rate), flow (< 100 Hz) or pressure control (<1000 Hz).

In practice, for slower processes, the anti-aliasing filter can often be a low order, as the main source of noise in industrial environments is known and related to the mains supply. This permits the hardware designer to take advantage of the features of analogue to digital converters (ADCs) which integrate the input signal over an interval of one mains period. This then gives a notch in the (sin(x)/x) frequency response matched exactly to the main noise source and all its harmonics (figure 1). For convenience of implementation, it is often desirable to fix the integration period exactly, but then the mains rejection degrades as the mains drifts from its nominal frequency - the rejection reduces to only 27dB if the noise frequency drifts from 50 to 48 Hz. Temperature controllers are typically designed to operate with 50 milliVolt of series-mode mains noise, while sensing thermocouple signals with sensitivities as low as 11 microVolt/°C, so mains rejections exceeding 70dB are required. To achieve this rejection, finite impulse response (FIR) filter techniques are applied, in addition to the hardware anti-aliasing filter.

Figure 1   Frequency response of integrating ADC

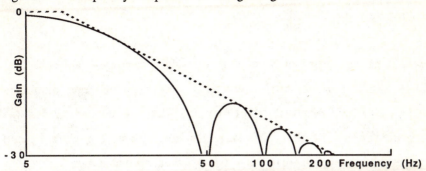

Some control domains - notably flow control - require further filtering of the input signal purely because the measurement technique itself can be noisy. In such cases the control engineer is given the ability to set a time constant for a single order filter on the input signal - this final version of the input signal is conventionally called the process value (PV). The disadvantage of this approach is that the PID algorithm is then attempting to control this filtered version of the input, which may require retuning the control parameters to account for this lag. This raises a significant point that the transfer function seen by the PID and tuning algorithms, includes the transfer functions of the input and output converters in the instrument as well as those of the process and its transducers. Consequently, design of these stages can either enhance or degrade overall control performance.

2.2. Amplitude resolution

Errors due to amplitude resolution are due mainly to cost constraints in the design of the instrumentation, although higher resolution often implies slower ADC rates. The requirement again varies with the application domain, being very dependent on the accuracy of the available sensors.

From a control point of view it is desirable to provide an extra four times resolution (2 bits) over that required for the displayed accuracy, since the PID algorithm is likely to hunt across the resolution limits of the ADC stage. In such cases, the extra control resolution may be achieved by using either normal process noise or artificially generated 'dither' added to the input signal, and feeding the control algorithm via an IIR input filter using a maths package of the required precision. This then generates samples with values which are effectively interpolated between those provided by the limits of the ADC resolution.

It is not unusual for process instrumentation manufacturers to sacrifice some ADC resolution, in order to achieve more flexibility in the instrument design. Thus if some applications may be limited to a 0 to 100.0 range on the sensor output, while others require 0 to 1000.0, the ADC is designed to provide the required resolution on the larger range, leaving the lower range with a lower percentage resolution. This is very typical in temperature control applications where users are just as likely to require 0.25°C resolution at 1200°C on one process, as 0.25 °C at 100°C on another.

Linearisation of the sensor's measurement characteristic is required both for display purposes, and in order to better achieve linear control requirements. This also interacts with resolution problem, as a sensor with a large gain variation across its characteristic can give much worsened effective resolution were the lineariser requires the highest gain.

Typically, the external sensors are the major source of error, (with accuracies specified down to 0.25%) and the requirement on the control instrumentation is more one of repeatability. This implies having very good gain and offset stability in the input hardware, under varying ambient conditions and over extended periods of time. Failure to achieve this adds further 'noise' on the input signal to the control algorithm, which most often cannot be removed by further signal processing.

3.   CONTROL ALGORITHMS

Digital implementations of PID algorithms most often treat the three terms independently and sum the result to give a control output which is then limited according to the actuator limits (figure 2). This section details the function and implementation of each term and the calculations resulting in a PID control algorithm (equations 1 and 2).

$$OP = 100/XP.(1 + 1/(s.Ti) + s.Td).ER \qquad (1)$$

where $\quad s.ER = d(ER)/dt$

or

$$OP(t) = OP(t-1) + K.ER(t) + L.ER(t-1) + M.ER(t-2) \qquad (2)$$

Figure 2   Block diagram of basic PID algorithm

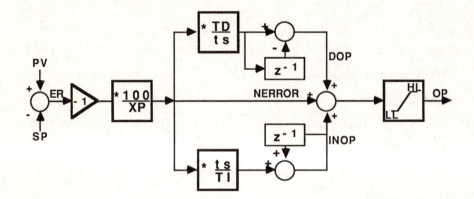

3.1. Proportional action

The controller is normally configured as 'reverse acting' which means that, under proportional action, the output decreases as the PV increases, in order to achieve minimum control error (figure 2). Direct acting control is required for processes where the prime actuator output must increase with increasing PV - typically increasing pumped coolant flow in refrigerating processes, or increasing outflow in a level control process.

The rate of change of controller output with respect to changes in PV is determined by the proportional gain, and since the output actuator necessarily has saturation limits (maximum and minimum output levels), this gain is often defined as the percentage of instrument span over which changes in PV give proportional changes in output, i.e the proportional band (XP) = 100/Gain (%).

For proportional-only action, a further parameter - manual reset - is used to provide an output offset such that the controller can provide the required output level with zero error.

<u>3.2. Integral action</u>

Integral action is required on the majority of industrial processes, since the controller is expected to achieve zero steady state control error under changing process conditions. The integral calculation is approximated in the control algorithm by adding each error sample, scaled by the proportional gain, into an accumulator. This approximation assumes that the error does not change significantly between samples (a trapezoidal approximation would be more accurate), but this is good enough for controlling loads requiring integral times as low as 4 ADC sample periods. Since the error is also scaled by (sample time/integral time), it is necessary to perform this addition using extended precision arithmetic (typically 24 bit), in order to attain close control with long integral times and low gains.

A significant problem with integral action is that when the error signal is large for a significant period of time, which occurs every time the loop experiences a large setpoint step, the integral accumulator 'winds-up' to a value much larger than the maximum possible control output. If this is not prevented, the integral continues to increase until the error signal goes to zero, and then the error must go negative to decrease the integral back to the appropriate value for steady state control. The result of this is overshoot on setpoint changes, which most industrial processes cannot tolerate. Individual control instrumentation manufacturers have adopted differing techniques to solve this problem - but all these techniques use a simplified limit model of the actuator as a basis for these solutions. A typical strategy would be to inhibit integral action when the demanded output exceeds the actuator limit.

Where this default action does not give satisfactory start-up performance, some manufacturers have adopted the cutback concept which permits the user to define the large signal response of the controlled process independently from the small signal response. The cutback value is set by the user and defines the error signal at which the control action transfers bumplessly from fixing the output at its limit to linear PID action. Hence cutback can be used to achieve either a faster start-up (with small cutback values) or an overshoot-free start-up (with large cutback values) according to the specific process needs. A disadvantage of using cutback is that if a noise transient or load disturbance is large enough to generate a control error which exceeds the cutback value then the integral value is preset to that required for start-up, probably resulting in a slower recovery from such a disturbance.

3.3. Derivative action

The main use of derivative is to achieve improved control stability by increasing the phase margin of the open-loop transfer function. The effect is to provide anticipatory action in the control output, which for many processes permits tighter control, and faster response to both setpoint changes and load disturbances.

However,where the process includes significant transport delay, the increase in gain at high frequencies due to derivative actually causes instability, so industrial controllers permit derivative to be turned off giving a PI controller instead.

Figure 3   Bode plots for PID algorithm with derivative filter

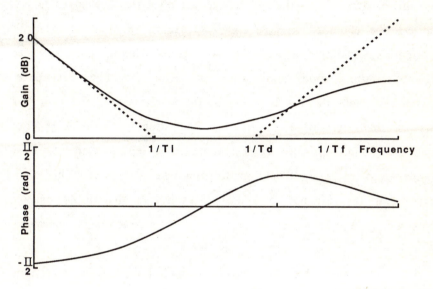

The increased gain at high frequencies is also a disadvantage for any processes having mechanical actuators, as the control action may be excessive, resulting in reduced actuator life. A further problem can occur through interaction with a badly designed integral desaturation algorithm. If this algorithm permits the integral output to be 'dumped' whenever the controller output exceeds its saturation limits, then this will happen when an input disturbance gets amplified by the derivative gain and all control action will be lost. Many PID algorithms employ an additional low pass filter linked to the derivative time constant, which limits the maximum increase in gain, but also limits the maximum phase advance provided by the derivative to much less than the (desired) maximum 90°. Figure 3 gives a Bode plot for a typical PID control algorithm with a low pass filter at 4 times the derivative corner frequency.

The use of derivative action to give a fast response to changes in PV can also be extended to give fast response to setpoint changes, by configuring derivative action on error signal. This is necessary for setpoint ramping applications, since the error should be essentially constant, whereas the PV is trying to follow the ramp. The result of this is that derivative acting on PV attempts to change the controller output to work against ramp following, and this is then compensated by integral action to minimise the error. This then delays both the response to a start of ramp and also to the end of the ramp, giving overshoot as the excess integral action is removed.

Derivative acting on the error signal does have a slight disadvantage, for small setpoint changes, as the resulting output kick due to derivative gives more tendency to overshoot. For many chemical processes instantaneous changes in control output are either dangerous or degrade the product, in which case derivative is not used at all, or it is only configured to act on the PV signal. In extreme cases, the controller may be required not to give even a proportional step in response to setpoint, in which case the controller output is 'debumped' to give a smooth transition to the new operating value.

3.4. Bumpless transfers

The action of 'debumping' the controller output during control action changes is very important to many industrial processes in order to maintain tight control of the process, throughout the various stages of operational control.

It is quite normal to set some processes up using manual control initially, and, once the process is close to its normal operating point, transfer to closed-loop (auto) control. In such cases, the controller output immediately after the change-over should be identical to the manual output demanded. This is achieved by forcing the integral output (effectively modifying the history of the error signal) at the instant of transfer, to balance the proportional and derivative outputs against the previous manual output.

i.e.

integral output = (previous - proportional - derivative) output   (3)

It is also often required to force the setpoint to track PV when in manual mode, so that the transfer to auto control is handled fully automatically.

This technique is also used to permit changes to control parameters to be achieved bumplessly - a feature that becomes most important in the context of self-tuning or 'term scheduled' controllers, otherwise the act of tuning to new conditions can actually result in increased disturbance to the process.

4.   OUTPUT CONVERSION

Most process controllers have basic output manipulation features such as a maximum limit and sometimes a minimum limit. These are most often used as backup safety measures - for example using the minimum limit to maintain a minimum gas flow on a burner, over the normal limits set by mechanical linkages. The output limits are then fed back to the PID algorithm to optimise the integral desaturation performance of the controller. A further safety limit offered on some process control instrumentation is an output rate limit which defines the maximum rate at which the output signal can slew. This prevents instantaneous step changes in control output, but, in consequence, adds very significant plant dynamics and requires handling by the integral desaturation routine.

Some temperature control processes require dual (heat-cool) outputs within a single controller, driving two independent actuators. This avoids the need for external split range output drivers, and thus results in much easier manipulation of the interdependencies of dual outputs. For a heat-cool controller, the cool output is treated as an extension of the heat proportional band for outputs less than zero. The gain in the cool proportional region can then be easily adjusted with respect to the heat proportional band, as can any offset.

The output is converted to suit the actuator on the process and is delivered either as a linear analogue signal or alternatively as a pulse width modulated (PWM) digital signal. In practice, linear analogue outputs are often generated by a fast PWM stream and then filtered by a low pass filter with a relatively high corner frequency. Two specialist digital outputs are detailed in the following sections, but a few general comments can be made on all the implementations.

It is not unusual for the output actuator to be non-linear - valves, for example, may be highly non-linear in their flow rate/position characteristic. Since this non-linearity may be very predictable, it is often pre-compensated by the inverse gain characteristic in the controller output converter - a very useful form of gain scheduling.

A characteristic of all digital to analogue conversion routines is that there is an associated frequency-dependent transfer function. In some cases the conversion may be performed at a much higher rate than the sample rate of the controller, and therefore the dynamics are irrelevant. Some implementations do, however, contribute significantly to the loop dynamics, and need to be included in any control design calculations.

4.1. Time-proportioned outputs

Pulsed outputs provide a convenient and efficient means of applying power to a process via electrical actuators, such as electrical heaters and solenoid valves in temperature processes. An important advantage is that the output is defined by the linear relationship between the demanded output and the 'on' pulse duration divided by the total (on + off) cycle time. This compares with the square law relationship which would be achieved by applying a linearly increasing voltage across a constant resistance heater.

A further advantage is that the application of power via a low voltage drop switch is highly efficient, although switch dissipation does become significant when sourcing very high load currents from semiconductor switches.

The problems associated with such output converters arise from constraints on the speed of operation. From a control viewpoint, the on plus off cycle time should be as short as possible in order to minimise consequent ripple on the process output, and also to minimise the contribution of this stage to the loop dynamics. However, mechanical actuators driven from time-proportioned outputs require as long a cycle time as possible in order to maximise their life, and hence minimise maintenance costs. Intelligent tuning algorithms have proved helpful here, in that these can be designed not only to tune the PID parameters but also to optimise the time-proportioning cycle time between these conflicting constraints.

Detailed analysis of the operation of these output converters indicates a division between those that give integral dynamics and those exhibiting derivative behaviour. The latter are most often preferred due to their fast response to changes in output demand from the control algorithm. This action is then typically constrained by a minimum pulse width algorithm, which prevents output 'glitching' and consequent reductions in actuator life, when the output of the PID algorithm is noisy. The application of minimum pulse durations results in small non-linearities in the output characteristic which are averaged out over time.

4.2. Motorised-valve positioners

Motorised valves are regularly used for controlling flows, and pose a specific control problem, which does not occur with valves driven by Electro-Pneumatic (E-P) converters. The valve opening (position) is changed by electrically driving its motor either backwards or forwards. The drive thus consists of raise or lower pulses, where the duration of the pulse determines the relative opening of the valve. Since the valve will also maintain its last position with no further input, it can be seen that these are characteristics typical of integral action. In addition to the normal valve characteristics of non-linearity and backlash, motorised valves also have three further disadvantages - integral action and minimum response time, and sometimes asymmetric action.

The integral action of the valve not only makes the loop more difficult to control, but also makes it more difficult to avoid integral wind-up in the PID algorithm, as the latter has no precise knowledge of when the valve is at its limit. The ideal controller output stage would exactly match the valve position to the output demanded by the PID stage - in control terms, therefore, this output stage must cancel the valve integral action with equal derivative action.

Figure 4   Block diagram of valve position controller

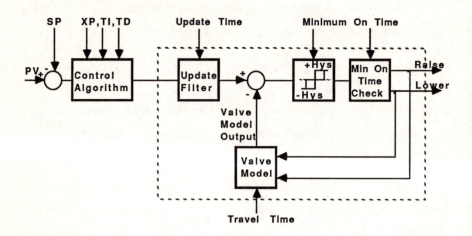

This is typically achieved by employing a model of the valve within the output converter, which is designed to follow the position of the controlled valve (figure 4). The PID output is continually compared with the output of this valve model, and if the PID output is greater than the valve model output, a raise pulse is output to the valve and the valve model output is raised by a corresponding amount. This continues until the valve model output is equal to, or just greater than, the PID output. Thereafter, if the output of the valve model is more than that demanded by the PID algorithm plus a given hysteresis, the opposite happens and a lower pulse is output to the real valve, and the model correspondingly reduced. In the hysteresis band neither output is active.

In order for the model to accurately track the actual valve output, the motor travel time (between valve fully closed and fully open) must be input to the algorithm as the integrator time constant of the model. This integrator in the feedback path then gives the output stage derivative action with the correct magnitude to cancel the dynamics of the real valve .

As for the time-proportioned output, the minimum pulse duration applied to the valve is constrained by a user selectable limit - the minimum on time - which effectively determines the resolution of the output. Valve wear is minimised by increasing this minimum on time parameter, but this then results in control hunting due to decreased output resolution. A better approach to reducing valve activity is to increase the time between PID updates to the output converter.

5.   TUNING TECHNIQUES

Although manual tuning techniques for PID controllers are well-defined for a wide range of processes [4], the application of these techniques is far from universal. This is often because the process engineer has little time to perform such basic tasks, and therefore may tolerate barely acceptable control performance. In other cases, the plant conditions may be continually varying and thus a single tuning will be insufficient.

For these reasons two basic approaches to automatic tuning are becoming more widely accepted - a one-shot autotune for commissioning, and a continuously adaptive tuner to cope with plant changes over time[2]. Process engineers are still rather cautious in applying continuously adaptive tuners, partly due to experience of failures in this area, but mainly due to a recognition that such a tuner can maintain good control despite degradations in parts of the plant, thus masking serious process problems. This highlights the potential use of such tools in the area of process fault diagnostics; an application which so far has had little exploitation in control instrumentation.

Most one-shot autotuners run an experiment on the process by applying a defined disturbance to the process, and monitoring the resulting process output. A convenient disturbance is that produced by replacing the PID control algorithm by a relay, giving one or two cycles of on-off control at the setpoint. This approach has the advantage of forcing the test duration to the minimum required to extract sufficient information from a completely unknown process, as the oscillation frequency is approximately that at which the process gives 180° phase lag. The disadvantage of this test is that the resulting oscillation may exceed the safe/desirable operating range of the process.

The range of continuous adaptive tuning techniques is too broad to detail in this paper[1], beyond discussing the solutions in the context of two further sub-groups. This division is between the term- or gain-scheduling approach, which relies on prior knowledge of process variations, and those approaches which attempt to identify the parameters of an internal process model (either explicitly or implicitly).

Continuously adaptive tuners have the advantage of not applying a test disturbance to the process, but the approaches using process identification techniques then have the problem of distinguishing between the response of the process and the effects of exogenous disturbances. For this reason, process engineers tend to favour gain scheduling tuning for safety-critical processes which require adaptive control action, despite the extra work involved in obtaining the initial tunings under a variety of process conditions.

6.   COMMUNICATIONS

The use of micro-processors in single-loop control instrumentation has increased both the accuracy and the accessibility of the data available on individual process loops. Significant design effort has been put into ensuring that data is not corrupted by the adverse electrical environment experienced by controllers in industrial applications. This data can then be collected via serial digital communication links and used either to provide more information on the process, or to integrate the individual loop into a hierarchical control system.

The availability of this data has led to the rapid adoption of supervisory, control and data acquisition (SCADA) systems, which offer the benefits of integrated process control schemes, which were previously only offered by direct digital control (DDC) systems. SCADA systems supervising one controller per loop offer the additional advantage of single-loop integrity permitting multi-loop processes to continue functioning despite the failure of one controller. This is most important for manufacturing processes with a high set-up cost or high cost of batch rejects.

Serial links are the standard for industrial control communications because they minimise cabling costs and can be electrically robust. The trade-off is that the overall information flow rate is reduced, so, with the increasing amounts of data available from each control point, communication bandwidths are increasing and controllers are expected to give fast response to the link. This highlights one of the effects that adding communications to single-loop controllers has had on the design of the instruments. Most digital control instrumentation was originally designed to give highest priority to the PID calculation, resulting in significant latency in response to communications via the link. Software architectures have had to be redesigned to give faster response to communications, and some controllers give highest priority to this, increasing the control sample interval (and adjusting the PID calculation) when communications overloads occur.

Communication links have also had effects on the design of the instrumentation hardware, since it is required both for safety and interference reasons that the link should be galvanically isolated from the input/output (I/O) system. In some control domains the I/O may be floating at mains potential, while in others (such as the control of volatile processes) it is imperative that the energy that passes back to the process via the I/O is strictly limited. It can be seen, therefore, that the interconnection of different parts of a process via the communication system must be through isolation barriers.

Finally, as communication speeds increase, there is a greater tendency to interconnect real-time events across the link, resulting in closer coupling between physically distributed control points. This can give significant performance advantages where the overall safety of the control system is not compromised, but in distributed systems where safety is critical redundant links must be used.

7. CONCLUSION

This paper has sought to illustrate that the control algorithm is in fact a small part of the total set of integrated algorithms and techniques which are necessary to make a robust industrial controller. This wealth of practical experience that has been incorporated in PID control instrumentation is perhaps the main reason that such controllers have widespread application even on processes for which they are non-optimal.

In particular, PID controllers apply a variety of techniques to handle process and actuator non-linearities, according to the individual application domain. PID controllers have been enhanced to deal with process interactions, and integrated as a basic building block in some very complex control schemes.

As energy and material efficiency becomes increasingly important to industry as a means of cost saving, the more complex control interactions resulting from feeding back waste energy, may be better handled by a multivariable control scheme which is not based on PID control blocks. This will be more likely to occur in 'greenfield' sites, where the cost benefits of more optimal control can be examined, rather than when retro-fitting new controls to old processes. A pre-requisite for this step, however, is better knowledge of the controlled process, which can only come about with the adoption of the new generation process modelling and identification tools using data from the SCADA systems now in use throughout industry.

It appears therefore that several technologies are all now being integrated to give the enhanced process knowledge that may yet take us beyond the PID controller.

REFERENCES:

1. Astrom, K.J. & Wittenmark, B. 1989: "Adaptive Schemes", Adaptive Control, Addison Wesley, 3-14.
2. Gawthrop, P.J. Nomikos, P.E. & Smith L.S. 1988: "Adaptive temperature control of industrial processes - a comparative study", IEE Control 88 conference publication 285, 59-64.
3. Shannon, C.E. 1949: "Communication in the presence of noise", Proc.I.R.E.,37,10-21
4. Ziegler, J.G. & Nichols, N.B. 1942: "Optimum settings for automatic controllers", Transactions of A.S.M.E., 759-768

Chapter 5

Generating sequences for control

R. M. Henry

The way in which switching and sequences are generated for practical control in industry are introduced and described. Methods have changed with the introduction of PLC's though programming and methodology are still related to the earlier methods used.

The extra facilities built into PLC's are also reviewed. Such facilities make them a new tool in their own right with the ability to compute and control on-line.

Large systems require much checking and testing. Software tools and methodologies to facilitate this are an important way of reducing project costs and demonstrating that the system will do what it should (and nothing that it shouldn't).

1.   INTRODUCTION

A common requirement in industry is that actions should always be performed in some order. For example, given two conveyors, the downstream one must always be started first - otherwise the upstream one would empty material onto a stationary conveyor. And should the downstream conveyor be stopped for any reason at all, then the upstream conveyor must stop as well - for the same reason as before.

The provision of simple interlocks or sequences such as the above often appears to be beneath the dignity of control engineers and university departments. It is a humble but very necessary task which needs to be done, and done well. Such systems ensure the continuity of the instrument air supply, provide flame failure detectors on all sorts of burners, close down conveyors when one is stopped in an emergency and generally ensure that what has to be done is done at the right time and in the right order, despite what buttons are pessed. There are many such tasks on any plant and the efficiency and safety depend on them being well done.

2.   HOW SEQUENCES USED TO BE GENERATED

Before the arrival of the programmable logic controller (PLC) there were two approaches to the provision of such systems. Either use discrete logic elements (AND, OR-gates etc), or use relays. The former corresponds closely with the experimental work you may have met when first introduced to logic and the design procedures for its implementation using a minimum number of components (truth tables, Kharnaught mapping, de Morgan's theorem, Boolean algebra etc).

There is an alternative technology for implementing switching logic: relays. The relay is a power amplifier with a binary valued output. It provides several pairs of contacts which open or close when the relay coil is energised. See figure 1. Contacts are designated 'Normally Open' - NO or 'Normally Closed' - NC. This refers to the state of the contacts when the relay coil is not energised. The contacts are described as 'clean' in the sense that they are electrically isolated from the power to the relay coil.

Wiring NO switches in series to power a relay coil imposes an AND condition, wiring them in parallel an OR condition. Conversely, NC switches in series remove power when any one of them opens - an OR condition whilst NC switches in parallel will only remove power when all are open - an AND condition.

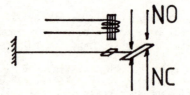

Fig.1   A simple relay

An example should make things clearer and ensure that you are thinking along the right lines.

Example

Four logic signals A, B, C and D are inputs to a selector box. The output will be one of these signals as indicated by two 'select' inputs. Give solutions in terms of both switching logic and relay logic.

This problem calls for something akin to a 4-way selector switch with the position selected by the two input wires. Whilst this is a useful conceptual model, neither solution looks anything like a switch!

2.1 Using switching logic

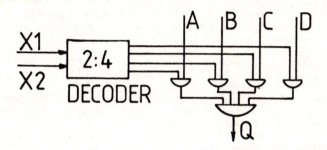

Fig.2 Selecting 1 of 4 signals by logic

2.2 Using relays

Here the select inputs, X1 and X2, and the signal input A - D are seen as switches which may be open or closed.

The select inputs are first bufferred so that their inverses are available. This allows us to decode the select inputs four ways, each being AND-gated with an input signal. Finally these four signals are OR-gated together.

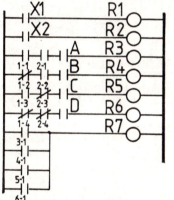

buffer input
signals

decode and AND
with A - D

OR together

Fig.3 Selecting 1 of 4 signals using relays

The state of relay R7 is the required output, Q.

The relay logic solution may look unfamiliar but it isn't too difficult to follow and corresponds closely with the switching logic solution. Note that relays R1 - R6 do not drive any output signals. Unlike R7, they are for internal use only.

2.3 Sequential logic

Without some sort of memory device one is limited to combinatorial logic and the sort of problem shown above. Using electronic logic units, memory is provided by the bistable or flip-flop. Exactly the same can be done with relays.

By wiring a relay as shown in figure 4 with a 'hold in' contact, it acts in much the same way as an RS flip-flop.

Pressing the NO (normally open) push button, PB2, causes the coil to energise and the NO relay contacts bridging PB2 maintain power to the coil even though PB2 re-opens.

Pressing the NC (normally closed) push button, PB1, de-energises the coil. The hold-in contact bridging PB2 opens (drops out) and the relay remains de-energised when PB1 re-closes.

The state of the coil reflects which button was pressed last.

This circuit has two stable states and is the 'flip-flop' of relay logic.

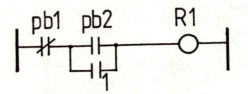

Fig.4 Relay with hold in contact

In fact, the cicuit shown in figure 4 is that used to start 3-phase motors in industry. The relay coil is a large solenoid which operates the three heavy duty pairs of NO contacts that carry the power to the motor itself. The hold-in contacts take the form of a more modest NO pair which handle the amp or two of current drawn by the starter (relay) coil.

2.4 Timing

In industry a fairly common requirement is timing. When you switch on your car ignition, the oil pressure warning light comes on but it signifies nothing because the engine is not running. You know that and ignore it except to note that the bulb is OK. If it were to come on whilst the engine were running, that would be a serious matter.

Much the same applies with sophisticated plant equipment such as compressors. Any lubrication failure must stop the compressor at once and raise an alarm. Wiring the NO oil failure pressure switch into the motor control circuit is easy enough but now the motor cannot be started because there is no oil pressure! We need to ignore the oil pressure switch for a few seconds at start-up to give the oil pressure time to build up, say 10 to 15 seconds. In the old days this would have been done by a timing relay.

The timing relay responds immediately to an input in one sense but delays when the input is in the opposite sense. There are four combinations:

1.	NO	closes immediately	times open
2.	NO	closes after delay	opens immediately
3.	NC	opens immediately	times closed
4.	NC	opens after delay	closes immediately

In many applications the requirements on accuracy and repeatability are low. The above example is just such a case. The delay is generated pneumatically and is adjustable in the range 1 second to 40 seconds or so.

2.5 Timing in ladder logic

In ladder logic there is no exact counterpart to the timing relay. Instead there is something more useful which looks rather like a resettable monostable. This block has two inputs, an input to trigger the delay and a reset. The latter can clear the timer whenever it is timing. The time is usually specified as integer seconds.

2.6 Counting

The PLC has counting facilities. The counter is rather like the timer but instead of being indexed by a clock, it is indexed by the external events being counted.

This gives rise to a need to handle BCD (binary coded decimal) arithmetic which is also supported. This allows an operator to set up input data on thumbwheel switches. In this way it would be possible to automate the transfer of a selected quantity of liquid from one tank to another with the PLC not only carrying out the operation but observing all the procedure and safety checks as well.

3. IMPLEMENTING LOGIC FUNCTIONS ON A PLC

Over the last 15 years, PLC's have come to be widely applied. Just how widely is shown by a recent survey (Control & Instrumentation Jan 1988) which lists more than 270 different devices from some 71 manufacturers. They range from very modest devices replacing a few relays to very large devices with a host of extra facilities. More about these later.

Clearly, in an introduction such as this it would not be appropriate to survey the field or to deal with one particular product from one manufacturer. What is intended is to give a general appreciation of the facilities to be found and the ways in which the software is organised.

3.1 Internal organisation

It is probably easier to start with the internal organisation. In an earlier example, the distinction was made between relays which had an output to the outside world and those relays which were for internal use only. In computing terms, a relay coil is represented by a storage location which contains 1 or 0 depending on whether the coil is powered or not. This state can be read as many times as we may want and is equivalent to a relay having as many contacts as we may want, both NO and NC. There is no need within a PLC to buffer as we may need to do in the real world when a relay has too few pairs of contacts. This is a useful simplification when designing. Similarly, we don't need to differentiate between sets of contacts as was the case when real relays were used.

To decide whether a 'relay coil' should be 'energised' or not we need to scan all its input switches. Some of these will be external to the PLC and others will be internal, relating to relays which may or may not drive outputs. Clearly, we need to specify what inputs are to be read/looked at and how they are interconnected to generate the AND and OR relationships required. This calls for some software.

3.2   The PLC's software

The software in a PLC differs from most software in that it forms an endless loop which scans the inputs and internal states and updates the outputs and internal states. Each cycle takes but a few milliseconds and the 'domino' effect as relays trigger sequentially on subsequent iterations is not a serious problem. After all, the same thing would happen if you were really using relays! If anything, the delay will be less. Sometimes, the programmer might make use of the certain knowledge of which 'relay' will trigger first in order to do something which would not be possible (because of indeterminancy) if using real relays.

This endless loop program is written by the manufacturer and it interprets the interconnections specified by the user. It is therefore an interpreter. In this sense it is like that better known interpreter, BASIC. The analogy applies further. Both the BASIC programmer and the logic programmer call what they write 'program'.

3.3   User programming

The problem now becomes one of designing what the user program should look like and creating the interpreter to bring this about. The two commonest solutions have already been briefly mentioned. The first is to interpret the Boolean expressions required. At first it seems the simplest and most obvious way but remember that we will be primarily interested in sequential operations.

The second approach is to program in ladder logic. This means specifying the solution in terms of the relays you are not using! This approach is widely used for a number of reasons. The first applications of PLC's were to replace existing relay applications, the projects being carried out by engineers familiar only with relays. It made excellent sense to drive the PLC's from the proven solution. In this way the new devices gained ready acceptance which in turn contributed to their rapid market penetration.

Whether one specifies the function in ladder logic or in boolean terms, the manufacturer written software is <u>interpreting</u> what the user has written. The user sees what he writes as 'program' but the interpreter has to turn this 'program' into action.

Clearly, we can choose any valid way of representing the logic/sequence so long as we write an appropriate interpreter. There will be very little difference in the hardware requirements of the PLC. Already there

is progress along this road with attempts at natural language programming and the use of nets or directed graphs in sequence specification.
The future would seem to lie with direct interpretation from the
specification and one such attempt has been reported (Dallas 1986).

4.   AIDS TO PROGRAM DEVELOPMENT

Getting programs together is a slow, error-prone business when checking
and documentation are taken into account. All help is gratefully
received. PLC's are no exception to this.

For small, simple systems, the programming may well be done with a plug-
in, hand-held keypad. Debugging might be done with a switch box to
simulate the external switch closures and check that the PLC provides
the right outputs under all conditions.

For larger systems, something more sophisticated is required. This could
well take the form of a personal computer communicating with the PLC
over a serial link. This would provide editing and emulation facilities
with eventual PROM firing or down-loading of the developed program. Such
systems would also make the documentation task much easier.

5.   PUTTING A PLC TO WORK

If you had a computing problem and you went to the nearest computer shop
and bought the first computer you liked the look of, you would rapidly
find you hadn't solved you problems. In fact they would probably be much
worse. The correct way of proceeding would be to consult an expert .

PLC's are just the same in this respect.

The first thing you need to do when you have found an expert to help you
is to tell him exactly what you want him to do for you. It sounds simple
enough. An example might help to show what I mean.

Consider a hopper which serves three bagging machines which weigh the
product into 50Kg bags. The bags are moved by a series of conveyors to a
pallet packer where twenty bags are made up and sheeted together. They
are then removed by a forklift truck either to a waiting lorry or to
storage. The idea is that all man-handling of the bags is eliminated.

The flow of bags to the pallet packer needs to be intermittent whilst we
would like to run the bagging machines continuously. This calls for some
buffer storage capability between bagging and packing and the idea is to
change the speed of the conveyors so that bags start to queue up when
there is a break in packing. Once packing resumes the queue is quickly
dealt with - at least that is what happens when everything operates
normally. Bagging only ceases when the queue had grown too long and
hopefully this will not happen frequently.

That is a superficial description and gives a general 'feel' for the
problem.

The detailed specification in words had to spell out exactly the
conditions under which each motor would start, stop, speed up or slow
down. It had to specify what should happen under every possible
eventuality. By PLC standards the application was not large or

especially complicated. However, you may be surprised to learn that the specification document was 3 cm thick, took more than two weeks to prepare and represented a cost of several thousand pounds. Nevertheless, the customer thought this was a worthwhile investment.

Such a document, known as a functional specification, is the basis of the agreement between the supplier and the client. That's what the customer had ordered, that's what the supplier is contracted to deliver. Cries of 'Oh! That's not quite what we wanted' will lead to greater expense and, worse still, to delay.

From such a document programming can commence. This is not a particularly large problem and could be dealt with by one programmer. Larger problems would involve several programmers and then the further complication is that the separate sections must all work together properly.

Writing in ladder logic has a few things in common with writing in assembler language. It is powerful and flexible but progress is not necessarily very rapid. Each programmer learns the standard tricks and many problems involve applying a few simple ideas many times over. A good example would be where one conveyor stops another. In such systems we can imagine the process a a lot of separate processes, each linked to the next in one straightforward way. So although there might be a lot of things to control, the problem is essential rather simple. In short, difficulty is not necessarily related to the number of things to be controlled.

For the more difficult sections a design methodology is a useful aid to getting everything right.

5.1   State transition matrix

Something much more difficult might be the control of a drink dispensing machine where a child might not press the buttons in the right order. Whatever happens, the machine has not got to 'hang up'. Here a technique which helps is the state transition matrix (STM). Down one side of a non-square matrix we show each 'state' and along the other side each possible input. In the matrix we show which next state a particular input will drive the system to from a given present state. If an input has no effect in some state then there will be no entry in the matrix. This is a clear concise specification, one can be certain that nothing has been overlooked and thnigs look good. However, there are a couple of difficulties which stop the technique from being widely applied. The first is that there can only be one state at a time. This is fine for many systems but when there are parallel subsystems it is often easier to think in terms of the states of these individual subsystems instead of being forced to distinguish each of a large number of conbinations as as a separate state.

The second limitation is that one could have the same input acting as a 'toggle' between two states. For example when switch J cause a change from state 5 to state 9, and also back from state 9 to state 5. The problem is that a latched relay is the equivalent of an RS flipflop which requires separate set and reset inputs. Ironically, this problem would not arise if one set out to program the STM in an ordinary programming language such as BASIC. What has happenned is that a limitation of relays has been faithfully copied into ladder logic programming. Fig.5 shows the STM for a stop watch with two buttons, A and B, and four states 'timing', 'stopped', 'zero(reset)' and 'lap'. There are two toggles in this STM.

STATE	INPUTS	
	A	B
Zero, Z	T	
Timing, T	S	L
Lap, L		T
Stopped, S	T	Z

Fig.5 State transition matrix for a stop watch

5.2   Sequencing and sequences

Sometimes this word is used in the sense 'ensuring tha things happen in the right order' as for example with the conveyors. At other times we might thing of a sequence as following directions such as a recipe for preparing a batch of material.

Sometimes the progression from one stage to the next is simply on the basis of time elapsed - a time driven sequence - but more often it is on the basis that the end of the current stage is recognised on some other basis (e.g. cook until golden brown). The latter is an event driven sequence.

Time driven sequences are quite easy to generate since all the different stages are driven off a single clock; things happen because it is time for them to happen. The event driven sequence is a little more complicated because we need feedback from the process about what is happenning.

5.3   Nets and the co-ordination of parallel subsequences

A further complication is that we may need to co-ordinate several event driven subsequences into one big sequence where different things are happenning at the same time but independently. At this stage we discover that natural language was a very good way of describing a serial event driven sequence (i.e. one where only one thing happens at a time) but has severe limitations when there are parallel operations. The way forward in such systems is to combine natural language with a directed graph or network. Such a representation belongs to a class of nets known as 'Petri nets'. Petri nets have been used to model a wide range of sequences and to even mention a few is beyond the scope of a chapter such as this. Suffice to say that a few years ago I felt that the way ahead in controlling such processes was to replace the ladder logic with a petri-net interpreter which could be driven directly from the specification. (Dallas 1986).

I still think the idea is a good one but I now feel that such an approach needs to be properly integrated with ladder logic (to do the easy stuff), state transition matrix and any other useful technique that may be developed. Just how this should be done is somewhat more problematical.

Tele-Mechanique, a French firm producing PLC's developed Grafcet, a petrinet like system for linking together blocks of ladder logic. Whilst this is short of direct interpolation is does avoid the problem I have just highlighted whilst still providing a useful service.

5.4   Software Engineering

There are a lot of ideas current in conventional software writing designed to produce much more 'workman-like' code. These include such ideas as structuring the programming, taking a top-down approach etc, etc. The same ideas are only now beginning to be applied to programming PLC's (Akerman 1989). There is a very long way to go. The use of macros or subroutines looks quite advanced to the PLC programmers.

An example might help to draw together some of the different ideas introduced in this section.

5.4.1   Two toy trains on the same track

The cost of equipment to demonstrate PLC's is iniquitous and amongst the best value for money is an 00gauge trainset. It also generates more interest amongst students than pneumatic pistons and microswitches.

Consider a single loop of track on which we want to run not one engine but two. Obviously, the problem is that one will go faster than the other and there will be a collision. To overcome this we will split the track into eight sections electrically isolated from each other (but still forming a loop. To detect the passage of an engine over a given point on the track we will plant reed switches between the rails and fix a small magnet under each engine. The switch will close for a moment as the engine passes over and it is recognised that there may be a problem of contact bounce. See Fig.6

The problem is straightforward enough to state. There needs to be a section of unpowered track between each of the engines. At the end of each track section an engine will be able to proceed providing the next section is available to be powered i.e. not immediately behind the other engine.

Being convinced of the need for a structured approach I started to look at this problem from the point of view of Petrinets and STM. Inspiration first came in terms of the intuitive solution. I sat down and wrote out the ladder logic directly. To do this I first looked at the single engine problem. Here the next track section is always available and the solution is as shown in Fig.7.

With two engines there is a possibility that the next section may not be available and the engine will be halted. This means that there is a need to 'remember' that the engine is waiting for power to be put onto a track section so we introduce a further eight relays to latch the closure of the switches nestling in between the track. The solution is now as shown in Fig.8.

Having solved the problem I next turned to how I thought I should have been able to solve it. Take STM first. There are 8 track sections where the first engine can be. The second engine cannot be on the same section, nor on the sections immediately in front or behind. That leaves

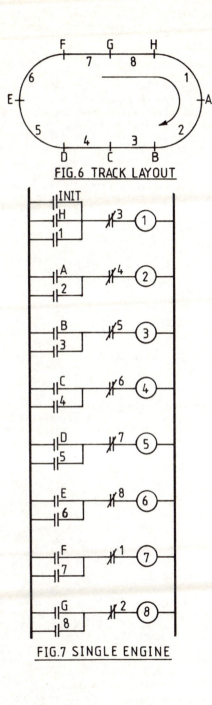

FIG.6 TRACK LAYOUT

FIG.7 SINGLE ENGINE

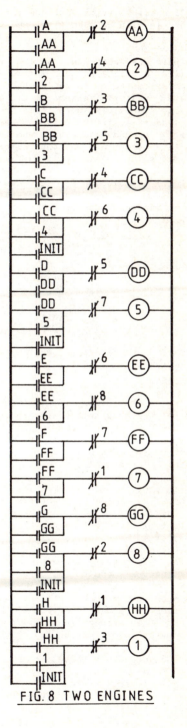

FIG. 8 TWO ENGINES

·only 5 possible track sections for the second engine. There is no need
to distinguish between first and second engines so there will. 8x5/2=20
possible states.

The petrinet approach ran into even more trouble   (Waire 1989) and
revealed a hitherto unremarked difficulty in applying petrinets to
switching control. In petrinets, meaning is only attached to a 'place'
(=state) that is 'marked' (i.e. active) and no meaning is attached to
unmarked places. In short, this is a limitation which does not exist for
the ladder logic solution since one can freely use either NO or NC
contacts to generate either meaning. It was for this reason that the
intuitive solution was the easiest to write down and why, when the
petrinet solution was achieved, it was twice as complicated.

The Petrinet solution for just one track section is shown in Fig.9. When
eight such diagrams are concatenated into a loop the whole thing looks
like a wonderful doodle but does not help anyone to understand what is
happenning.

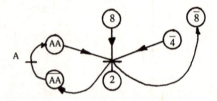

Fig.9 Petri net controlling track section 2

5.5 Testing

Testing in the factory is a lot cheaper than testing on site so a system
will be put through its paces in the factory using lights as outputs and
switches as inputs. Making all the connections and then putting the
system through its paces to detect errors in the coding takes a
considerable length of time. It also introduces the possibility of
connection errors whilst testing.

One thought towards shorter testing times is to use the serial line to
the PLC by which it is programmed or has its program down-loaded.
(Crosland 1989) In this way one should be able to sit at a PC and do all
the testing without leaving one's desk. It's an attractive idea, even
recognising that there is still a need to test the i/o modules
themselves since they were part of the 'lights and switches' testing
program. Such systems are commercially available - but only at a steep
price that only the professionals can afford.

Even when you've forked out the asking price of £10,000 or so, you have
only tackled part of the problem. It is important to look at the system
as a whole and recognise that we need to look at all the stages that go
between the specification and the finally commissioned system.

Some thoughts along these lines include:

1.   Eliminating sources of human error by using a computer to list
     which outputs and internal states are affected by each input.

2. Which inputs and which internal states affect each output (the converse of (1) above.
3. Methods of partitioning big problems into smaller ones.
4. Application of appropriate methodologies to different partitions
5. Improved methods for ensuring that a trick used repeatedly is copied by the computer so that the scope for human errors is eliminated.

An area where help may be forthcoming is the testing of LSI logic circuits. This is an area where testing has been built into the design of the chips and it could be that we should be adding things to the ladder logic to may testing more efficient.

From this very cursory treatment I hope you will be able to appreciate that there is a large problem area which is costing indutry a geat deal of money each year, and that the problems are indeed worthy of academic consideration.

6.   THE ADVANTAGES OF USING PLC's

PLC's have gained wide acceptance in industry for a number of reasons but the best one is that they save money. In addition, they have other advantages such as reliability, easier checking, shorter installation times etc.

In looking at cost, one is comparing a small box of microelectronics made in large numbers with the low tech solution of individual relays wired and checked by a skilled electrician. The latter is slow and labour intensive so installation times and commissioning times can be reduced with considerable cost savings. Until quite recently relays might have been considered for small scale 'one-off' applications. Now they are queueing up for a place in the Science Museum! As part of the control system for a piece of sophisticated plant made in large numbers the PLC scores very heavily and became the preferred way of doing things some time ago. The program can be copied and checked very easily and replaces the whole of the internal wiring required by the relay system. Any subsequent updates will only require a new PROM rather than re-wiring.

So far we have concentrated on designing and building the system. PLC's also have advantages when it comes to maintenance. Individual relays are quite reliable but as numbers rise, reliability must fall. Faults cost money, are a skilled man will need to spend time tracing the fault at great expense. PLC's are much more reliable. When one does go wrong it can be readily replaced and the program transferred.

7.   OTHER PLC FUNCTIONS

Up till now, I have presented the idea of a PLC as simply replacing a cabinet full of relays and timing relays. That is not strictly the case. Whenever one re-engineers a product using micro-electronics, one rarely produces an exact copy. Each technology has its own constraints and the change gives one the opportunity to throw off old constraints. You have already met the timer and the counter which do not have direct counterparts in terms of hard-wired relays.

Let us recall that although the PLC is busy pretending to be a lot of relays, it is really a computer. It can do all the usual things that computers can do as well. In particular it can 'talk' to other computers over a serial line. It could act as a remote control device relaying data back to a central control room with consequent savings in cabling costs and installation time. It can also perform numerical functions and continuous control. Its timing is really accurate unlike that of the old timing relays which were a bit 'hit and miss'.

From the above you will not be surprised to learn that all the suppliers of large scale distributed systems offer a PLC device capable of sitting alongside the rest of their system on a data highway, usually for its batch control capabilities.

The PLC market has grown very quickly and the manufacturers have added features which extend beyond switching control. For those wanting an analogue i/o loop as well as digital i/o some offer continuous control as well.

7.1 Continuous control

Several manufacturers now offer PLC's capable of controlling a handful of continuous loops. Texas were among the first with their PM550 system now superceded by the PM550C which offers 16 loops of PID control.

The need arose when automating batch processes where most of the effort lay in the switching and in ensuring that the correct operating procedures are always observed. With a PLC going in to look after this duty, it seemed rather silly to use a separate controller for each of the analogue loops, especially as the PLC already contained a processor.

The conventional analogue controller be it pneumatic, electronic or digital, comes with an operator interface, only we don't often think of the facilities it offers along these lines. Obviously, it has to be set up with the right settings and in some cases there would never be any need to look at the controller again. In such cases there would be no need for an operator interface. Practice has shown that this is rarely the case and the conventional controller offers the following features:

1. Display setpoint
2. Display measured variable
3. Display control signal (output)
4. Display controller settings
5. Display whether direct or indirect acting
6. Allow manual/automatic changeover
7. Allow manipulation of control signal when in manual

Some of these facilities will be reserved to the engineers.

In recent years, single loop controllers have been redesigned around microelectronics and it is interesting to note that they still provide all these facilities although differently accessed and with the enhanced features such as a choice of algorithm and connection to a serial data highway.

The PLC's offerring continuous control may provide a less flexible interface. For example, controller settings may require a question and answer session using a hand-held keypad plugged into the PLC. Displays may not be to such a high standard. However, if the PLC is being driven from a data highway, then the information can be relayed back to a central control room for operator presentation using some form of colour graphics display.

8.   DON'T THESE PLC'S HAVE ANY DISADVANTAGES?

Yes they do. Whenever you think you've solved a problem in technology, you find you've simply created another problem. Take the motor car for example. Solving the personal transport problem creates traffic problems.

The first problem concerns safety. If a PLC is to be incorporated in a safety system then questions need to be asked about the system failure modes. Using relays, these were very simple. Using PLC's things are not so clear and approval has to be sought from an engineer who says 'you've convinced me this is more reliable than relays, but how can I be really sure you haven't made a mistake in the programming?' The Health & Saftey Executive are currently looking at this in their report on Programmable Electronic Systems (PES).

This problem will be partly resolved by better design methodologies and direct interpretation. In fact, by software engineering.

One problem still remains and that concerns the failure modes of PLC's. They may only breakdown rarely but when they do, what might happen? Can we detect it? Can we protect ourselves against it? This is a very tricky question; I don't know the answer. Perhaps we might avoid it by moving to systems with inbuilt redundancy and self-checking.

9.   WHERE NEXT FOR PLC'S?

Fifteen to twenty years ago PLC's were the product no-one even knew they wanted. Now no-one would think of doing things without them. In the face of such relentless progress don't expect me to predict the future. However, certain trends do seem fairly clear.

The first is to larger scale applications. PLC's removed a limit on complexity imposed by relays and discrete logic, they have not yet reached their own hardware limits.

Software development is a major cost in large-scale applications and better methodologies are required for designing such systems. These should allow easier specification and documentation. They may also lead to direct interpretation. The latter step would be very worthwhile indeed. Coupled with this will be better methods of testing the user written software to prove its integrity.

Already applications outstrip the ability of single PLC's and there is a need to interconnect many such devices in large scale applications. To do this a special serial bus has been defined with a protocol which guarantees the worst case access time to the bus. This is MAP, the Manufacturing Automation Protocol developed by General Motors in the USA and now adopted as the IEE802 series of standards. This allows different manufacturers' equipment to be interconnected on the same bus, something not previously possible. The jargon phrase for this is 'Open System

Interconnect' or OSI. It seems destined to have wide application, certainly wider than the systems for which it was originally designed. Presently, the industry is awaiting the development of the chips which will make the OSI an affordable reality. This was promised for tomorrow (writing in February 1987). Rewriting in December 1989 I can confidently say we are 22 months nearer the solution - but it is still tomorrow as far as realistic costs are concerned.

10.   CONCLUSIONS

PLC's have quietly revolutionised the way some things are done. It has all happenned in quite a short time and there is something rather satisfying about seeing the high technology being put to good work.

Such rapid progress and a lack of academic involvement had meant that the software engineering aspects of PLC's are only now being properly addressed. They would seem to have much to offer in terms of improved reliability and reduced programming and commissioning times.

Ten years ago batch process control was rarely mentioned. It was the 'Cinderella' of the process industries. The large continuous plants were only possible because of the control systems fitted. They were expensive and could only be paid for from the benefits of large scale production and so a symbiosis had developed which left batch production in the cold. Many new products such as those developed by biotechnology demand batch processing. These are high value, low volume products and PLC's offer possibilities for their low cost automation. Now everyone is looking at 'Cinderella' in a new light!

11.   BIBLIOGRAPHY

1.   Control & Instrumentation Vol 20 no. 1 Jan 1988 - Special issue on programmable controllers including a survey (updated each year).

2.   Dallas P D 'Computer control of continuous and batch processes using a Petri-net driven interpreter' PhD thesis, Bradford, Dec 1986.

3.   PMT604 Real-time Control - Open University post-experience course, first published 1987.

4.   Henry R M & M Webb 'Ladder Logic for se quence generation - A Methodology' Meas & Control Vol 21 no 1 Feb 1988.

5.   Henry R M & B Abu Bakar 'A methodology for implementing a state transition matrix on a PLC' Meas & Control Vol 22 no 1 Feb 1989.

6.   Akerman J 'Structured programming for today's PLC's', Control Systems, Nov 1989.

7.   Crossland A 'Testing software for Programmable Logic Controllers' Open University T401 project, Sept 1989.

Chapter 6

Real-time computer networking

G. C. Barney

ABSTRACT

An instrumentation system or a control system does not have to exist at one specific location nor does a system have to exist in a specific cabinet. Nowadays the idea of centralised systems is replaced by the concept of distributed systems. Control engineers long ago discovered the frailty of centralised control rooms and took the advent of the microprocessor to distribute control actions to the point of application. Instrumentation engineers working in the same epoch seized the opportunity to incorporate intelligence into the point of measurement. And computer engineers developed the technique of distributed computing by means of networking.

This chapter investigates the means of communication between instruments and controllers using digital computers. It deals first with the ideas of distributed systems and discusses serial interfaces. Later the modern techniques of local and wide area networking is applied but not in detail.

1. DISTRIBUTED CONTROL SYSTEMS

A distributed control system aims to control a widely disperse plant. The distribution of control also necessitates distributed instrumentation (sensors and actuators).

Consider Figure 1, which depicts a hierarchy of systems.

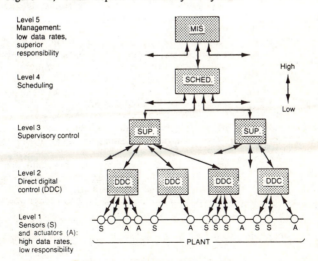

Figure 1 A hierarchy of systems

Level 1   This lowest level connects directly to a plant or process and incorporates the measuring sensors and controlling actuators. These devices may incorporate microprocessors or custom chips to deal with A/D–D/A conversion, calibration, linearisation, actuator operation, and so on.

Level 2   This level uses microprocessors to provide closed loop control to a plant or signal processing for an instrument. In the control case Direct Digital Control (DDC) loops may be involved with complex three term (PID) control algorithms or, in the instrument case, a fast Fourier transform (FFT) may be evoked.

Level 3   The "instructions" level 2 requires to specify its operating modes, data transfer, activity, etc., which can be provided by a supervisory computer, which may be a powerful microcomputer or minicomputer. In the control context this level provides supervisory control of the set points into the level 2 DDC loops.

Higher levels   The extension of the concept of levels can be carried indefinitly to many higher levels. The figure shows two more as an illustration. Level 4 is shown as a scheduling computer where decisions are made to optimise a factory or refine output according to (say) economic demands. This machine will be a minicomputer. The next level (level 5) could be a corporate mainframe machine at a remote location, where a management information service is provided.

A feature of a hierarchical system is that the lowest levels have high data rates to and from the connected process but a low decision responsibility, whereas at the higher levels responsibility is severe and data sparse.

Example 1
What are the actual advantages of distributed control systems?
Some of the advantages of distributed systems are given in Table 1.

Table 1 Some advantages of distributed control systems.

FEATURE	COMMENTS
Communication	Communication between computers is facilitated by a digital format using standard procedures
Cost	Serial transmission is cheaper than parallel connections or analogue wiring. The use of specialist integrated circuits is cheaper than relay or hard wired logic.
Reliability	Greater reliability is possible owing to the large numbers of processors, which can operate, at a degraded level, in the event of communications failure or individual processor failures.
Data integrity	Data can be checked for errors.
Improved performance	More complex control strategies can be implemented or improved signal processing incorporated.
Modularity	It is easier to separate functions into logical modules. This provides for clean interfaces between modules. Allows easier testing and modifications.

2   INTERCONNECTING DEVICES

As can be seen from Section 1, efficient and effective communication between levels and within levels is of paramount importance to the operation of distributed systems. Communication is a concept for the interconnection of devices. Table 2 gives the main forms of communication. This chapter is not concerned with parallel data transmission or with Connection Type (c).

Table 2 Interconnecting Devices

FEATURE	DESCRIPTION
Connection type:	(a) Computer and Computer
	(b) Computer and Equipment
	(c) Equipment and Equipment
Distance:	(a) Locally
	(b) Remotely
Transmission:	(a) Parallel
	(b) Serial
Mode:	(a) Asynchronous
	(b) Synchronous
Quantity:	(a) Character
	(b) Bulk

Three types of communication are described in detail in the following sections:

(A) Local, serial, asynchronous, character: computer to equipment.
(B) Local, serial, synchronous, character/bulk: computer to equipment.
(C) Local/remote serial synchronous bulk: computer to computer.

3   SERIAL COMPUTER AND EQUIPMENT CONNECTIONS

3.1 The terminal port

Every digital computer possesses a terminal port (or is capable of possessing one), which enables terminals (teletypewriters, visual display units, graphic terminals, etc.) to be connected. Figure 2 shows how a terminal (T) can be connected directly or remotely to a computer (C). When distances increase (over 300m) amplifying devices called modems are employed. These allow data transmission in analogue form, often by frequency shift keying, to be carried over telephone networks. The modem achieves this as it functions as a modulator/demodulator. Simpler devices to achieve the same purpose are called short haul modems or line drivers.

Transmission between computer and terminal is serial, asynchronous by character and synchronous by bit. Illustrations of the transmission of a string of characters and a single character are given in Figure 3(a) and (b) respectively.

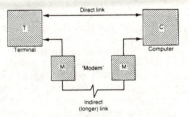

Figure 2 Terminal port connections

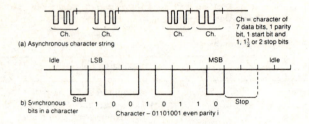

Figure 3 character strings

The characteristics of a computer asynchronous port for terminal/computer connections are well established as listed in Table 3. It should be explained that half duplex uses the same pair of wires for both directions of transmission so that simultaneous input and output which is allowed with full duplex operation is not possible. Non responsive signalling implies that there is no acknowledging signal provided back to the sender whenever data is received.

Table 3 Asynchronous Port Characteristics

Hardware connections:

Local loop	20 mA
EIA(USA)RS232/442/443	± 3 V
CCITT(Europe)V24	± 6 V

Transmission speeds (bits per second)
    75, 110, 150, 300, 600, 1200, 1800, 2400, 4800, 9600
    (Output/input speeds may be different, e.g. 1200/75)

Transmission mode:
    Full Duplex   FDX
    Half Duplex   HDX

Signalling protocol:
    Non-responsive
    Responsive

3.2 Character coding

Figure 3b shows each transmitted character comprises 8 data bits plus synchronising bits. One of these (bit 8) is a simple error checking bit termed the   bit. It is used to make the total number of '1's in the character always odd or even depending on whether even or odd parity is required. The seven bits are assigned as alphabetic, numeric, punctuation, communication or terminal control characters, see Figure 4. These character assignments are known as US-ASCII (American Standard Code for Information Interchange) and have been accepted world-wide.

Examination of the table shows that columns 3-8 define printing characters except for delete (DEL). Columns 1 and 2 define control characters. The characters carriage return (CR), linefeed (LF), bell (BEL) etc. control the operation of the connected terminal and the remaining characters, start of heading (SOH), end of text (ETX), acknowledge (ACK), etc. control the operation of a connected modem device, if any. Full explanations of all these control characters can be found in programming manuals.

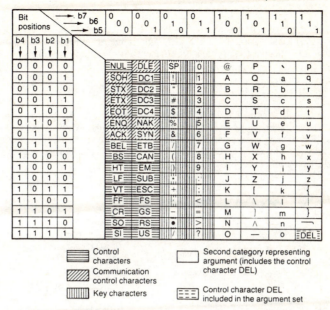

Control characters

Communication control characters

Key characters

Second category representing argument (includes the control character DEL)

Control character DEL included in the argument set

Figure 4 Character assignments according to US–ASCII

3.3 Asynchronous equipment connection

It is possible to connect equipment other than terminals to a computer by means of the asynchronous terminal port. For example, data from an 8 bit A/D converter could be sent as an ASCII character. The computer may, however, interpret a bit pattern 1000 0011 end of text (ETX) as a command to terminate the transaction. To avoid this, the computer can be put into   mode, where it will simply pass all characters through "without, comment". This is dangerous as the transmitting device may fail to restore the computer to character mode after a transmission. An alternative is "character stuffing" where characters in columns 1 and 2 are preceded by a special defining character such as ESC (escape). In the event, ESC, ETX is received the computer accepts the bit pattern ETX as data, but if ETX only is sent it is accepted as a genuine command. Probably the safest mode of operation at the expense of an increased overhead is to send only six bits of data by permanently setting bit 7 to '1', thus avoiding the problem.

3.4 Commentary

The use of a terminal port for equipment connection is a popular procedure. Its advantages are:

- (a)   most digital computers have at least one port,
- (b)   a software driver probably exists to handle the I–O,
- (c)   the port characteristics are well defined and
- (d)   integrated circuit (IC) hardware is readily available;

e.g. universal synchronous/asynchronous receiver–transmitter (USART).

Disadvantages are:

- (a)   loss of port,
- (b)   the computer may not be able to provide the transmission speeds required, e.g. 1200/75 or full duplex operation
- (c)   the transmission code may be restrictive, e.g. ASCII.

The method described is frequently used as it is regarded as "easy". However, each implementation is a "special case" with no standard solution, and time is wasted "reinventing the wheel". Clearly this technique is not an ideal solution.

4   COMMUNICATION STANDARDS

The distributed control system shown in Figure   1 depicts a large number of computer to computer communication paths. These paths require: point to point links, multipoint links and network connections. The traffic carried includes: interactive program activity, file transfers and remote resource access. To communicate effectively between devices a must be established.

Example   2
Using the ASCII code, illustrate how a simple protocol may be devised.
The solution can be seen in Figure 5.

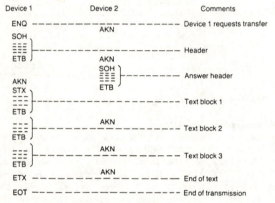

Figure 5 A simple protocol

Device 1 sends device 2 a header indicating the amount of data, block sizes, odd or even parity, other error checks, etc. Device 2 repeats the header to ensure correct receipt. Then device 1 sends 3 text blocks and receives acknowledgements. Finally an ETX, EOT ends the exchange.
Example   3 illustrates a simple solution which only takes care of a data exchange. It is necessary to establish a standard protocol.   The International Standards Organisation (ISO) have proposed such a protocol in its Open Systems Interconnection (OSI) 7-layer model (another hierarchy!).

4.1 The ISO seven-layer model
The layered approach to networking stems from operating system design, namely to break down an ultimate goal into a number of modules, which are then fashioned to meet a specific function. Modules have two specific forms of relationship:

* *interfaces* - relationships between different modules in the same node

* *protocols* - relationships between equivalent modules in different nodes.

Typically for a hierarchical system a higher module will receive a service from a lower module in the same node. Protocols establish message formats and rules for their exchange. Figure   6 depicts the 7-layer protocol. The philosophy and standards which make up each level are complex and of a highly specialist nature. Fortunately for the users it is possible for this sophistication to be hidden. Table 4 indicates and describes the seven layers.

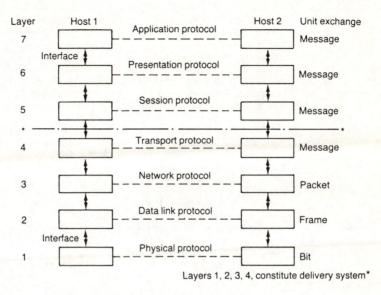

Figure 6 ISO seven layer model

5   LOCAL AREA NETWORKS

Wide Area Networks (WANs) have existed for some time (the US ARPANET is one example) but did not necessarily conform to the ISO model. They are defined as communication networks over a wide area geographically and internationally at moderate transmission rates (kilobits/s). Local Area Networks (LANs) were inspired by the trend to distributed computing and are gaining acceptance for distributed control (e.g. PROWAY). Their definition is "a communication network which covers a limited geographical area at high speed". Generally this is accepted to mean 1–5km, 1–100Mbits/s with error rates 1:10.

5.1 LAN Topologies
A local area network has a number of primary attributes:

(a)   inexpensive transmission media and modems (twisted pair, TV coaxial cable, fibre optics, etc.).

(b)   high data rates up to 100Mbit/s (ideally 140Mbits/s for TV picture transmission).

(c)   a high degree of interconnectivity (potential to communicate with any other node).

(d)   no control node (there may be an administrative node).

(e)   each node listens to each transmission and accepts those addressed to it.

To comply with the above there are a number of leading LAN topologies: star, ring, bus, tree, as illustrated for eight nodes in Figure 7.

Table   4 The ISO 7-layer model

LAYER	DESCRIPTION	STANDARDS
Physical	Defines the electrical and mechanical interfacing to a physical medium. Sets up, maintains and disconnects physical links. Includes hardware (I–O ports, modems, communication lines, etc.) and software (device drivers)	RS232–C RS442/443/449 V.24/V.28, V.10/V.11. X.21,X.21bis, X.26, X.27, X.25 level 1
Data link	Establishes error free paths over physical channel, frames messages, Error detection and correction. Manages access to and use of channels. Ensure proper sequence of transmitted data	ANSI–ADCCP ISO–HDLC LAP DEC DDCMP IBM SDLC,BISYNC X.25 level 2
Network	Addresses and routes messages Sets up communication paths Flow control	USA DOD–IP X25,X75 (e.g. Tymnet, Telenet, Transpace, ARPANET,PSS)
Transport	Provides end to end control of a communication session Allows processes to exchange data reliably	USA DOD–TCP IBM SNA DEC DNA
Session	Establishes and controls node system dependent aspects. Interfaces transport level to logical functions in node operating system.	)))))
Presentation	Allows encoded data transmitted via communications path to be presented in suitable formats for user manipulation	) FTP)))) JTMP)
~Application (user)	Allows a user service to be supported e.g. resource sharing, file transfers, remote file access, DBM, etc.	)) FAM))

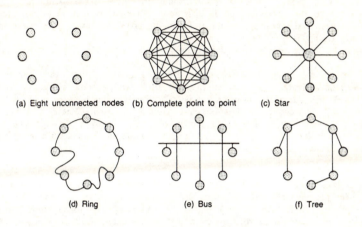

(a) Eight unconnected nodes (b) Complete point to point (c) Star

(d) Ring   (e) Bus   (f) Tree

Figure 7 LAN topologies

Complete point to point (Figure 17.7(b))
Here all nodes are connected directly to the remaining seven nodes. This is inefficient, expensive and difficult to operate.

Star (Figure 17.7(c))
This is a common topology as many sites already possess a star communications network from control room, central computer or telephone exchange. The disadvantage is that the loss of the central switch will bring down the network.

Ring (Figure 7(d))
All the nodes are now connected to the nearest adjacent nodes in a notional ring (it may not be symmetrical). Each node must repeat signals received and problems arise if a node fails or the ring is broken or shorted out.

Bus (Figure 7(e))
In this topology all nodes are attached to a common bus. This topology is more robust should the bus be broken or shorted. It is susceptible to a node "sticking–on" and continuing to transmit, thus shutting out all other activity.

Tree (Figure 7(f))
This is a variation of the bus, often used to reduce overall bus traffic by connecting nodes who frequently communicate on one bus and connecting the sub buses together by means of a gateway.

5.2 Access Methods
Section 5.1 described how devices may be connected to a LAN it is now necesary to consider how they may access the network.

Star LANs
When a device wishes to talk to another device it sets up a call to the central switch, which will route the data through if it is free and the other device is free. The access is therefore controlled by the central switch.

Ring LANs
Messages on a ring pass in one direction round the ring. Each node takes each message, demodulates it, buffers it and if it is not addressed to that node remodulates it and passes it on. If the message is for the node it takes a copy, marks the original as accepted and passes it on.
Rings may be configured in several ways to allow access:

(a) *Tokens* A unique token is passed round the ring and any node may remove it, transmit a message and add the token to the end. A receiving node will mark its safe acceptance, which will cause the transmitting node to remove the message.
(b) *Message slot* A sequence of bits circulate around the ring. Where the slots are of fixed length a node, which detects an empty slot, may insert a message and mark the slot full. If the slots are not of a fixed length a node may transmit a message to fill the space available.

Bus LANs
Access to a bus LAN may be synchronous (token passing) or asynchronous (contention). In the former case a token is passed from node to node by some predetermined route (similar to a ring). Only the owner of the token may transmit. This is a slow procedure so asynchronous access is common. Here any node which wishes to transmit listens to the bus and if it is not busy transmits its message. It can happen that two nodes transmit at the same or nearly the same time (nodes can be quite far apart and might not hear a node, which has previously started transmitting owing to the propagation delay) in which case a collision will occur. To prevent this an access control mechanism called Carrier Sense Multiple Access/Collision Detection (CSMA/CD) is used to resolve the contention.

5.3 Comparison
A star network based on the X25 Standard at layer 3 is simple to implement but operates very slowly (9.6kb/s–48kb/s). As these protocols have been used in WANs for some time a good body of experience exists. It is also possible by the use of PADs (Packet Assembler Disassembler) To connect asynchronous devices as terminal lookalikes. Standards exist for this connection (X3,X28,X29). Example configuration: Packet Switching Exchanges (PSE) in Britain.
Ring networks cannot transmit onto an idle bus unless they have the token or can see an empty slot. The loss of the token or the message slot results in the ring crashing. Difficulties exist in adding and removing nodes whilst service is maintained. Example: Cambridge Ring, Britain.
Bus topologies offer more promise with considerable activity with Ethernet (Digital–Intel–Xerox), PROWAY (Purdue University and BSI) and the IEEE 802 standard which is a combination of Ethernet and PROWAY. It is difficult to separate out errors and collisions with long buses although the CSMA/CD algorithm is simple to implement. A bus operating a token passing scheme can avoid collisions and provide a guaranteed access time important in real time work (PROWAY). Examples exist in Britain and the USA).

6   LANs IN A CONTROL ENVIRONMENT

The discussion of communications has been at a superficial level. This is deliberate as users of a LAN to link equipment and computers together should not need detailed technical knowledge of the service they are taking, but only of the interface they see. A second reason for the superficial treatment is that LAN technology at all layers is developing rapidly. Hence, for example, a detailed description of CSMA/CD is inappropriate as better algorithms will probably emerge.

Table 5 A layered process control application

LAYER	FUNCTIONS

Application/user system:

Application	Management Functions Supervisory Control Sequence Control Direct Digital Control Specific tasks: Motor Control Temperature measurement A/D conversion
Presentation	Operator function services: Displays, Printers Control panels
Session	Real time operating system functions Device drivers Interrupt handling Application modules

Delivery system:

Transport	Translation of process name to address. Local routing within same node. Remote routing of messages. End to End error checks. Priority structure.
Network	Error detection of link or node failure. Provision of alternative routes.
Data-link	Provides error free communications over physical channel. Error detection CRC, LEC, etc. Access control Address control.
Physical	Bit/character detection. Modulation/demodulation Physical connection/disconnection Simple error detection

Example 3
Describe the functions which will be performed by the various layers when a LAN is applied to process control. The response to this example is tabulated in Table 5.

BIBLIOGRAPHY
Bibbero, R.J., "Microprocessors in Instruments and Control", John Wiley, 1977.

BSI "Draft standard specification for PROWAY for distributed process control systems", 80/32741DC–65A, 1981.

Digital, "Introduction to Local Area Networks", Digital Equipment Corporation, Maynard, 1982.

DIX (Digital,Intel,Xerox), "The Ethernet: a local area network" specification, 1980.150/TC97/SC16N227: "Reference model of open systems interconnection", 1979.

Liu, M.T., "Distributed Loop Computer Networks", Advances in Computers, Vol. 17, pp.163–221, 1978.

"Papers on Networking", Computer Networks, Vol. 4, pp.197–283, 1980.

ACKNOWLEDGEMENT
This is an abridged version of Chapter 17 from "Intelligent Instrumentation" by G.C.Barney, Published by Prentice Hall International, who have kindly given permission for its reproduction here.

Chapter 7

Parallel processing for computer control

G. S. Virk

1 Introduction

Ever since the first general purpose microprocessors became available on the market in the 1970s, control engineers have recognised the tremendous benefits of direct digital control (DDC) techniques, and have consequently adopted their use in many wide and varied applications. These early microprocessors were rather primitive; they were 8 bit machines working with clock rates of the order of 1MHz and possessed little software support - usually the programming had to be performed in the assembly language. The situation is somewhat better now. The capabilities of electronic chip manufacturers have evolved to the stage where very large scale integration (VLSI) is possible and 32 bit microprocessors working at 25 MHz are commonly available; such modern microprocessors are powerful computing devices, which are able to perform many millions of instructions every second. In addition, they can also have on-chip floating-point support and are able to be programmed in standard high-level languages.

Although significant advances in computing speed have already been made, and the processing power available to users has grown at an outstanding rate, it is unlikely that this rate can be maintained for much longer because of physical limitations, namely, that signals cannot travel faster than the speed of light (3×10^8m/sec). To illustrate the difficulty posed by this restriction, assume that an electronic device can perform 10^{12} operations per second. Then it takes longer for a signal to travel between two such devices one-half of a millimetre apart than it takes for either of them to process it. In other words, all the gains in speed obtained by building superfast electronic components are lost while one component is waiting to receive a signal from another. The problem cannot be solved by packing components closer and closer because, firstly, there is a physical dimensional limitation of the etching process, and secondly, unwanted interactions can occur if the discrete devices are too close, causing a reduction in the reliability.

These physical limitations are unfortunate because they mean that the processing demands from the users, which are still increasing, will not be achievable. The difficulties have been noticed and alternative approaches are being considered, and a major development in this respect, is to insert parallelism into computer systems. Traditionally computer systems have consisted of a single central processing unit (CPU) that is used to execute all the instructions. and so all the data and commands pass through the CPU in a serial manner, giving rise to the Von Neumann bottleneck encountered in systems of this type. By adding more processors into the systems, parallelism can be introduced and the computations speeded up by being performed on a distributed processing network.

Many types of parallel computers can be considered, depending on the configuration chosen, but they can be grouped into the following three categories (see Sharp [2]):

1. Multiple instruction stream, single data stream (MISD).

2. Single instruction stream, multiple data stream (SIMD).

3. Multiple instruction stream, multiple data stream (MIMD).

These are shown more clearly in Fig 1, where the traditional single processor type computer is also shown as the SISD machine.

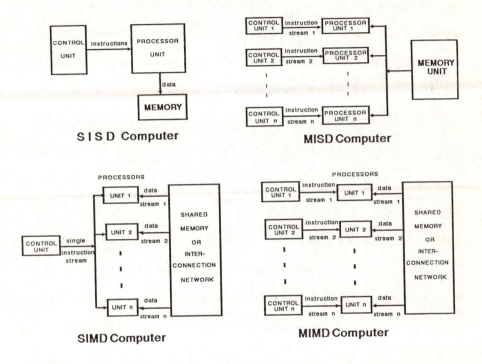

Figure 1: Computer Systems

Clearly, the parallel computers have several processors executing instructions on some data. The MISD form of computer systems have a single data stream which is operated upon by the various processors in some defined way. Hence parallelism here is achieved by letting each processor perform, in general, a different operation on the same data provided from a shared memory system. In contrast to this is the SIMD form of computer where all the processors operate under the control of a single instruction stream, but upon different data streams. The processors here operate synchronously that is, they all execute the same instruction but each on a different datum; there are global clocks available that ensure a lock-step operation.

For the general problems that need to be solved on an SIMD computer, it is a requirement that the processors be able to communicate amongst themselves during the execution of a program so that intermediate results can be exchanged. This can be accomplished either through a shared memory system and/or via an interconnection network allowing point-to-point connections between the processors.

MIMD systems are the most general, and most powerful, class of parallel computers. Here each processor is of the type used in SIMD computers except that each has its own control unit which can issue a different instruction. Hence all the processors are potentially executing different programs on different data, while solving various sub-problems of an overall problem. The technique is also known as functional processing, since each processor node performs, in general, a different function and so operates in an asynchronous fashion. As with SIMD computers, communication between the processors can be performed through a shared memory and/or an interconnection network.

It is worth mentioning that shared memory systems can give rise to memory contention problems, where adding extra processors does not give improved performance because the processors have to wait (idle) while access to memory is being achieved. An interconnected processor system, where each processor has its own local memory, does not suffer from this problem, and adding extra processors gives further improvements, provided the communication overheads are not excessively increased. With the interconnected architecture systems, it is obvious that various configurations as shown in Fig 2 can be employed depending on the application, the number of processors available and the precise computations, required to be performed, see for example Akl [8],

Decegama [9], and Bertsekas and Tsitsiklis [10].

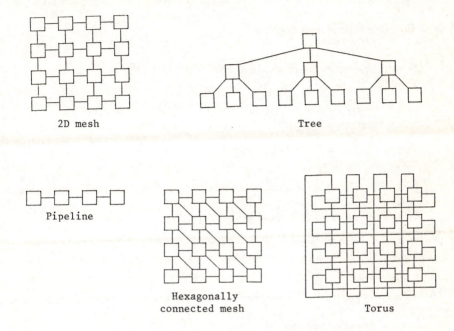

2D mesh   Tree

Pipeline

Hexagonally
connected mesh   Torus

Figure 2: Parallel Processing Architectures

A recent development in the area of parallel processing is the transputer (see INMOS

[3], [4], [5]), which allows MIMD computer systems to be implemented in a straight-

forward manner. The transputer is in fact a family of VLSI devices, each with a CPU,

on-chip memory, an external memory interface and communication links for direct

point to point connection to other transputers. By using these external links parallel

processing networks of transputers can be built up, and, with proper synchronisation

and communication, they can be used in a variety of applications, (see for example ref-

erences [11] - [13]). INMOS have developed a precise programming language, Occam,

for the transputer so that parallel constructs and synchronised communication can be

accommodated. A brief description of the language is now provided.

2   Occam Overview

The model of concurrency supported in hardware by transputers is the Occam model, which has the following main features:

1. Everything is made up of processes which exist in parallel.

2. Processes may be created, may die and may create other processes.

3. Communication between the processes is through channels.

4. A channel is a unidirectional, unbuffered link between just two processes and provides synchronised communication.

5. A process is either a primitive process or a collection of processes.

6. A collection specifies both its extent (what processes are in the collection), and how the collection behaves.

A typical Occam process model is shown in Fig 3 where process P1 is a primitive process whereas processes P2 and P3 are collections of communicating processes. Communication channels between the processes are also shown.

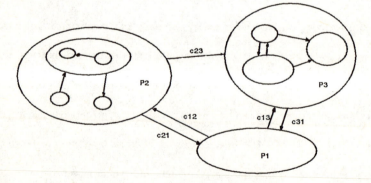

Figure 3: Occam Process Model

Programming in Occam is centred around the use of three primitive processes which are:

1.	Assignment:	x := 20
2.	Input on a channel:	chan.in ? x
3.	Ouput on a channel:	chan.out ! x

More complex processes can be formed by using these elemental commands with constructors, such as **SEQ** (sequential execution of component processes), **PAR** (parallel execution), **ALT** (alternative execution dependent on the first process to become activated), **IF** (conditional execution) and **WHILE** (repetitive execution while satisfied). The communication channels can be defined to accommodate various types of variables/data, for example, integer, real and byte. The channels provide synchronised communications between the processes and so data is transferred only when both the processes are ready. Further details regarding Occam can be found in the Occam 2 Reference Manual [6] . An Occam programming environment called the Transputer Development System (TDS) has been provided by INMOS, see INMOS [7], where full details of the system can be found.

3   Occam and Transputer Implementation

The transputer has been designed to efficiently implement the Occam model of concurrency. Fig 4 shows a three transputer system implementing the Occam process model of Fig 3, with the inter-transputer communications being made possible by use of the external transputer links. Although, in the following discussion, we will concentrate on programming in Occam, it is possible to program transputer hardware in other high level languages such as Parallel C and Parallel Fortran.

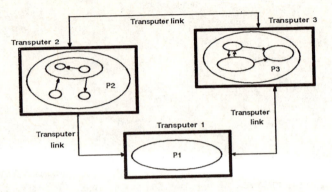

Figure 4: Transputer System Model

A transputer has its own task schedular, and so it can be used to run a number of concurrent processes at the same time by time-slicing the processor. Hence the communication between processes can be through the external transputer links, if the processes are being executed on separate transputers, or through internal software channels if the processes are being run on the same transputer. Each transputer link provides two Occam channels; one is an input channel and the other an output channel - both have to be specified in the program. The external link channels between different transputers and the internal software channels on the same transputer are seen to be identical at run time. Therefore it is possible to develop a solution to a problem independently of the actual network configuration upon which it will be finally executed. That is, the solution could be developed on, say, one transputer, and once a working solution had been obtained the various processes can be allocated to different transputers and the appropriate channels to transputer links for final implementation. The Occam facilities that allow this are as follows:

1. **PLACED PAR** - parallel execution of processes as before, but now the processes are placed on separate transputers.

2. **PROCESSOR** number **transputer.type** - identifies a **number** to a processor and what **transputer.type** the processor is so that the configurer can check that the correct compiler was used (T2 for the T212 transputer, T4 for the T414 transputer and T8 for the T800 transputer).

3. **PLACE** channel.name **AT** link.address - allocates a channel identified as **channel.name** to a link identified as **link.address**. The **link.address** also identifies whether the channel is a source or a destination. For example, for the 4 links on the T414 transputer the 8 Occam channels (4 input and 4 output) are identified as follows: Link 0 gives channels 0 and 4, Link 1 gives channels 1 and 5, etc, where 0,1,2,3 are output channels and 4,5,6,7 are input channels.

Having given a brief overview of some of the software and hardware aspects of transputer systems, a control example to illustrate these techniques will be presented.

4   Example: Control of flexible structure systems

In this section we will use parallel processing methods to distribute the various tasks (processes), in a control application onto a network of transputers. Such an approach may be useful in areas where the computational burden is large and/or fast cycle times are required, which cannot be achieved using conventional sequential computing techniques.

In a general real-time control system design the main tasks to be performed may be categorised into the following headings:

1. Modelling.

2. System simulation for model validation.

3. Control design, implementation and performance assessment.

4. User interface.

Since each heading can, in principle, require vast computational resources it is important to have the flexibility of being able to design the appropriate computer architecture so that the execution time is not excessive and real-time performance is possible. We will present a design using the above functional processing approach for the control of flexible structure systems, (see Virk and Kourmoulis [17], [18]). In particular, we shall be considering the problem of vibration suppression in structures such as aircraft wings, helicopter rotors and solar panels on satellites. These structures can be investigated by considering the problem of suppressing the vibrations of a cantilever beam in flexure as shown in Fig 5.

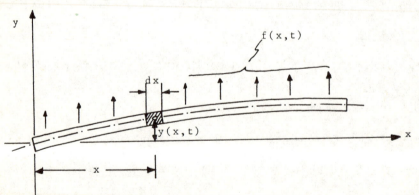

Figure 5: Cantilever System in Flexure

It is well know (see for example Timoshenko [15]), that the transverse displacement at any point x along the beam, when subjected to a distributed force $f(x, t)$, at time t, is given by $y(x, t)$ which is the solution of the partial differential equation (PDE):

$$\mu \frac{\partial^4 y(x, t)}{\partial x^4} + \frac{\partial^2 y(x, t)}{\partial t^2} = \frac{1}{m} f(x, t) \tag{1}$$

where μ is a constant depending on the physical material properties of the beam, and m is the mass per unit length of the beam. Cantilever systems have an infinite number of modes and therefore require significant computing resources to simulate and/or control adequately.

The real-time control of such structures can be tackled by using a reduced-order model to obtain a control law that can be implemented on the cantilever system itself. In our work the cantilever system did not exist in reality and was therefore simulated by numerically solving equation (1). The network of transputers, as shown in Fig 6,

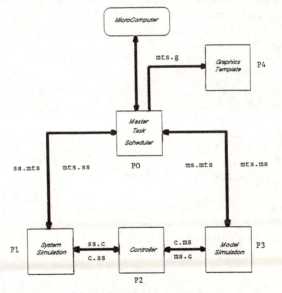

Figure 6: Block Diagram of Computer Configuration

can be used to solve this problem, where each process block can be a single transputer or a network of transputers connected in a suitable configuration, as discussed be-low, to achieve the real-time performance sought. The system Occam configuration file for blocks each containing a single T800 transputer could take the form shown in Fig 7.

```
{{{ System Configuration
    CHAN OF ANY chan.in, chan.out, mts.g, ss.mts, mts.ss, ss.c, c.ss:
    CHAN OF ANY c.ms, ms.c, mts.ms, ms.mts:
    PLACED PAR
        PROCESSOR 0 T8
        PLACE chan.out AT 0
        PLACE mts.g AT 1
        PLACE mts.ms AT 2
        PLACE mts.ss AT 3
        PLACE chan.in AT 4
        PLACE ms.mts AT 6
        PLACE ss.mts AT 7
        Master (chan.out, mts.g, mts.ms, mts.ss, chan.in, ms.mts, ss.mts)
    PROCESSOR 1 T8
        PLACE ss.mts AT 0
        PLACE ss.c AT 1
        PLACE mts.ss AT 4
        PLACE c.ss AT 5
        System simulation (ss.mts, ss.c, mts.ss, c.ss)
    PROCESSOR 2 T8
        PLACE c.ms AT 1
        PLACE c.ss AT 3
        PLACE ms.c AT 5
        PLACE ss.c AT 7
        Controller (c.ms, c.ss, ms.c, ss.c)
    PROCESSOR 3 T8
        PLACE ms.mts AT 0
        PLACE ms.c AT 3
        PLACE mts.ms AT 4
        PLACE c.ms AT 7
        Model simulation (ms.mts, ms.c, mts.ms, c.ms)
    PROCESSOR 4 T4
        PLACE mts.g AT 7
        Graphics (mts.g)
}}}
```

Figure 7: Configuration file for Architecture Considered

We assume that the configuration file is preceded by folds containing the separately compiled codes (SC's) for each process. This is the highest level fold which is shown in Fig 8, where the location of the system configuration fold is also shown.

```
{{{ F Cantilever.example
... SC Master
... SC System simulation
... SC Controller
... SC Graphics
... System Configuration
}}}
```

Figure 8: Occam Program Structure

The three fold dots indicate that information is hidden away in a "fold" and can be seen by entering the fold, (see the TDS User Guide [7]).

The computer hardware that was available to implement the above architecture was a Meiko computing surface comprising, 17 T800 and 2 T414 transputers with a total memory of 13 Mbytes and a processing capacity of 275 MIPS. Hence further partitioning of the processing necessary for computing the control solution, as discussed below, could be done to speed up the execution time. The three main blocks/processes in the architecture, namely the system simulation, modelling and the controller can contain several transputers. These blocks are still inter-connected by transputer links/communication channels, as shown in Fig 6, and are still controlled by the master processor, which also drives the graphics hardware for the user-display. Each process proceeds in synchronised discrete steps, with each step producing a numerical result which is perceptible to the whole network. A brief description of each of these blocks will now be given.

4.1   Master Task Scheduler (P0)

This contains a T800 transputer connected to the host computer, and so, all access to the keyboard and screen is through this processor. As the name implies, it is used to initiate the program execution and to schedule and control the other processes.

4.2   System Simulation (P1)

The solution of the 4th order partial differential equation (1) exists in a space-time grid

shown in Fig 9. This grid originates from the finite difference approximations technique

(see for example Ames [19], and Collatz [20]), which can be used to express equation

(1) in the first central finite difference form as

$$y_{i+1,j} = 2y_{i,j} - y_{i-1,j} - \lambda^2 \{y_{i,j-2} - 4y_{i,j-1} + 6y_{i,j} - 4y_{i,j+1} + y_{i,j+2}\}$$

$$+ \frac{(\Delta t)^2}{m} f(x,t) \tag{2}$$

where

$$\lambda^2 = \frac{3}{4} \frac{(\Delta t)^2}{(\Delta x)^4} \mu \tag{3}$$

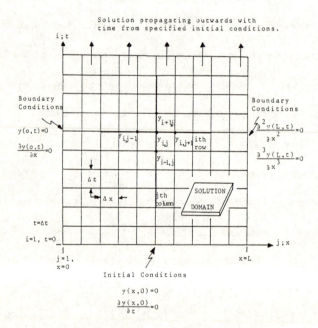

Figure 9: Solution Domain for PDE

The numerical solution then follows by using this equation together with known

initial conditions to give $y(x,t)$ at the chosen number of node points in a row by row progression satisfying the boundary conditions at each iteration.

These solutions may be computed on a single processor or the grid may be split in the vertical direction so that each region can be assigned a separate processor. The beam was partitioned into 20 sections (or "stations") and the 22 stations (20 actually on the beam and 2 for the boundary conditions) were divided into 4 groups, and each group assigned to a transputer using the configuration shown in Fig 10 to achieve real-time performance.

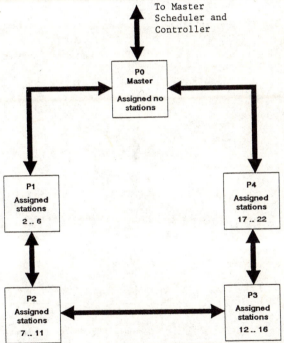

Figure 10: System Simulation Block

4.3   Modelling

The solution, $y(x,t)$, of the 4th order PDE that describes the motion of the cantilever beam in transverse vibration can, using the separation of variables method (see references [14] - [16]), be expressed in the infinite modal sum

$$y(x,t) = \sum_{i=1}^{\infty} \Phi_i(x)q_i(t) \tag{4}$$

where the spatial and time dependence of y has been separated into two distinct functions Φ_i and q_i for each mode. It turns out that in most applications only the first few modes are important, and so selecting the first m modes, gives a reduced-order description for the cantilever. Substituting the Φ_i and q_i into equation (1) results in a simpler form which can be written in the standard state-space form. The first five modes were used to produce a 10th order state-space representation with $x_6 = \dot{x}_1, x_7 = \dot{x}_2, \ldots, x_{10} = \dot{x}_5$, (see Kourmoulis [18]). This can be numerically solved in parallel on a network of 5 transputers using the architecture shown in Fig 11.

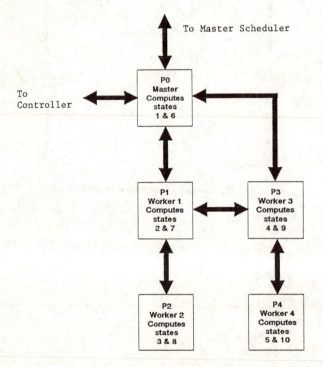

Figure 11: **Model Simulation Block**

4.4   Controller Design

Various design methodologies can be considered for obtaining a suitable control law. The approach taken was to design an optimal control law that minimised a linear quadratic performance index for the 10th order state-space model.

A parallel off-line algorithm was implemented to solve the set of Riccati equations and the results stored for on-line use. The Riccati equations can be solved in real-time if required, see Tahir and Virk [22] . The resulting control law is applied to the system simulation block and model simulation block, the results are shown in Fig 12.

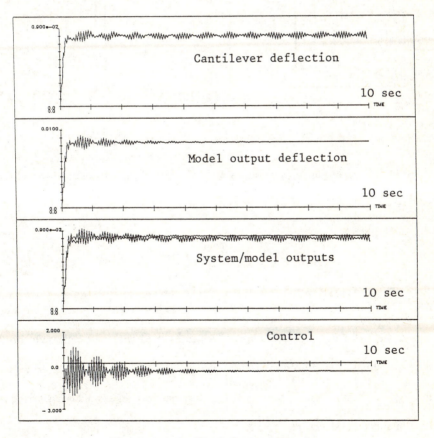

Figure 12: Optimal Control Results

The complete block diagram of the transputer network is shown in Fig 13.

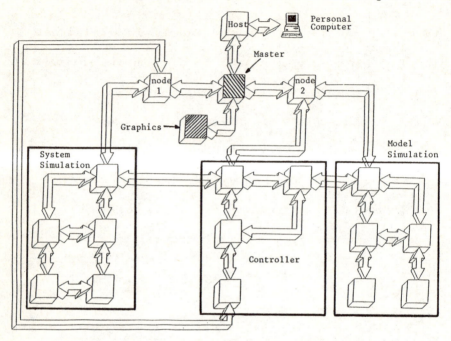

Figure 13: Complete Transputer Network System

A simulation time of 60 seconds took 40.82 seconds on this network and so real-time results are possible. The same problem on a Sun 3/50 workstation, coded in Fortran, took 1587 seconds.

5   Conclusion

The concept of parallel processing has been introduced with particular reference to control systems. The implementation of such techniques using transputer arrays programmed in Occam has also been presented together with an example considering the control of flexible structure systems.

References

[1] D I Jones and P J Flemming, Real-Time Control using Occam and Transputers, IEE Workshop on Parallel Processing and Control: the Transputer and other Architectures, University College of North Wales, Bangor. Digest No.1988/95, 1988.

[2] J A Sharp, An Introduction to Distributed and Parallel Processing, Blackwell Scientific Publications, 1987.

[3] INMOS Ltd, Transputer Instruction Set, Prentice-Hall, 1988.

[4] INMOS Ltd, Transputer Reference Manual, Prentice-Hall, 1988.

[5] INMOS Ltd, Transputer Technical Notes, Prentice-Hall, 1989.

[6] INMOS Ltd, Occam 2 Reference Manual, Prentice-Hall, 1988.

[7] INMOS Ltd, Transputer Development System, Prentice-Hall, 1988.

[8] S G Akl, The Design and Analysis of Parallel Algorithms, Prentice Hall, 1989.

[9] A L Decegama, The Technology of Parallel Processing, Vol 1, Prentice-Hall, 1989.

[10] D P Bertsekas and J N Tsitsiklis, Parallel and Distributed Computation: Numerical Methods, Prentice-Hall, 1989.

[11] IEE Colloquium on, The Transputer: Applications and Case Studies, Digest No 1986/91, 1986.

[12] IEE Colloquium on Recent Advances in Parallel Processing for Control, Digest No 1988/94, 1988.

[13] IEE Colloquium on Transputers for Image Processing Applications, Digest No 1989/22, 1989.

[14] Meirovitch L, Elements of Vibration Analysis, McGraw-Hill, 1986.

[15] Timoshenko S P, Young D W, and Weaver W, Vibration Problems in Engineering, John Wiley, 1974.

[16] Inman D J, Vibration with Control, Measurement, and Stability, Prentice-Hall, 1989.

[17] Virk G S, and Kourmoulis K K, On the Simulation of Systems Governed by Partial Differential Equations, IEE Conference Control 88, Oxford, pp 318-321, 1988.

[18] Kourmoulis K K, and Virk G S, Parallel Processing in the Simulation of Flexible Structures, IEE Colloquiium on Recent Advances in Parallel Processing for Control, Digest No. 1988/94, 1988.

[19] Ames W F, Numerical Methods for Partial Differential Equations, Academic Press, 1977.

[20] Collatz L, The Numerical Treatment of Differentiual Equations, Spinger-Verlag, 1955.

[21] Kourmoulis K K, Parallel Processing in the Control of Flexible Structure Systems, PhD Thesis (to be submitted), University of Sheffield, 1990.

[22] Tahir J M, and Virk G S, A Real-Time Distributed Algorithm for an Aircraft Longitudinal Optimal Auto-pilot (accepted for publication), Concurrency: Practice and Experience, 1990.

Chapter 8

Design of software for real-time system

S. Bennett

1. INTRODUCTION

A real-time computer system typically has the following characteristics:

Multiple inputs which may be:

*   discrete events occurring at any time;
*   continuously changing data values;
*   discrete data values.

Multiple outputs which may be:

*   discrete events;
*   continuously changing data values
*   discrete data values.

Timing: the actions have to be performed within the time scale set by the environment or synchronously with the environment. Formally the requirement can be expressed as:

*   the order of computation is determined by the passage of time or by events external to the computer;
*   the results of the particular calculation may depend upon the value of some variable 'time' at the instance of execution of the calculations;
*   the correct operation of the software depends on the time taken to carry out the computations

System state: the actions performed on the inputs may change according to:

*   the internal state of the program;
*   the state of the environment.

and

*   the inputs may change the internal state.

Figure 1 represents one possible model of a real-time system. In this model of a real-time system the outputs are function of the inputs and the internal state but the internal state can be changed by the state of the environment and the environment can be changed by the internal state.

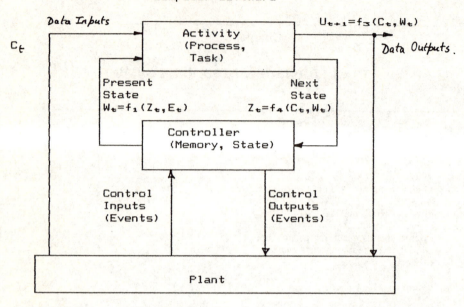

Computer Software

C_t

Data Inputs

$U_{t+1} = f_3(C_t, W_t)$

Data Outputs.

Activity
(Process,
Task)

Present
State
$W_t = f_1(Z_t, E_t)$

Next
State
$Z_t = f_4(C_t, W_t)$

Controller
(Memory, State)

Control
Inputs
(Events)

Control
Outputs
(Events)

Plant

External Environment

Figure 1. Finite State Machine Model

Figure 2 shows an alternative functional representation in which the transfer function is assumed to contain time and state dependent elements.

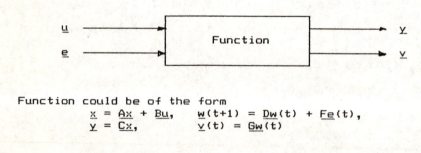

$\underline{u}$

$\underline{e}$

Function

$\underline{y}$

$\underline{v}$

Function could be of the form
$$\underline{x} = \underline{A}\underline{x} + \underline{B}\underline{u}, \quad \underline{w}(t+1) = \underline{D}\underline{w}(t) + \underline{F}\underline{e}(t),$$
$$\underline{y} = \underline{C}\underline{x}, \quad \underline{v}(t) = \underline{G}\underline{w}(t)$$

Figure 2. Functional Model

1.1 Classes of real-time systems

We can divide real-time computer systems into two categories:

(i)   Hard time constraint: the computation must be completed within a specified maximum time on each and every occasion.

(ii)   Soft time constraint: the system must have a mean execution time measured over a defined time interval which is lower than a specified maximum.

The first category - the hard time constraint - is obviously a much more severe constraint on the performance of the system than the second. The so called embedded systems i.e. systems in which the computer or computers form an integral part of some machine usually come into this category.

1.1.1 Hard time constraint systems

Some examples of hard time constraint systems are:

*   Control systems with the computer in the feedback loop: the control algorithm is designed for a specific sampling rate and if the computation is not carried out at this rate then the calculated control value will be wrong.

*   Computer operated alarm systems: the specification will normally require that on each and every occasion the computer responds to an alarm occurring within a specified minimum time.

*   Recording of events taking place in real time. If two events occur and the computer has failed to register the first event before the second one occurs, then only one event will be recorded. A minimum separation time between events will be specified and will form the maximum allowable response time for the real-time computer system.

1.1.2 Soft time constraints

Soft time constraint systems are typically interactive systems: an example is an automatic bank teller - the specification may require that over a 24 hour period the systems responds to a customer within an average time of, say, 30 seconds and on no occasion is the delay longer than 1 minute.

1.2 Task - a definition

We define a task as a segment of code which is treated by the system software (for example an operating system) as a program unit that can be started, stopped, delayed, suspended, resumed, and interrupted. A task is assumed to be capable of being executed concurrently, or pseudo-concurrently, with other tasks. Pseudo-concurrency is the running of several tasks on a single cpu. A task may be thought of as being equivalent to a process in the sense in which the term is used in computer science literature.

1.3 Program classification

Two important heuristic rules for designing real-time software are:

(i)   partition the software into two units - one containing activities which have to be carried out in real time, the other containing non real-time activities

(ii)   minimize the number of activities in the real-time partition.

These rules arise from a recognition of the differing levels of difficulty of constructing the different types of computer programs.   Empirical studies have shown that the design and implementation of real-time software is considerably more difficult and costly than is the case for non-real-time software.   Pyle (1979) drawing on the work of Wirth (1977) presented three categories of programming and these are discussed below.

1.3.1 Sequential

In classical sequential programming actions are strictly ordered as at time sequence: the behaviour of the program depends only on the effects of the individual actions and their order.   The time taken to perform the action is not of consequence.   Validation of the program requires two kinds of argument: (i) that a particular statement defines a stated action; and (ii) that the various program structures produce a stated sequence of events.

1.3.2 Multi-tasking

A multi-task program differs from the classical sequential program in that the actions performed are not necessarily disjoint in time: several actions may be performed in parallel.   Validation of multi-task programs requires the arguments used for the validation of sequential programs plus consideration of synchronization.   The tasks may be validated independently only if the resources used by each task (variables, devices etc.) are distinct and independent.   Sharing of variables or other resources implies the possibility of concurrency and makes the effect of the program unpredictable unless some rule (or rules) governing the synchronization of the tasks is introduced.

The introduction of synchronization rules is sufficient to ensure that the time taken to carry out individual actions does not affect the validity of the program.   Each individual task can proceed at its own speed: the correctness of operation depends on the correctness of the synchronization rules.

1.3.3 Real-time

A real-time program differs from the two previous types in that: (a)   certain of its actions may have to be performed within time constraints set by the environment; and (b) the sequence of its actions may be changed by events occurring in the environment. Inter-task synchronization rules do not provide a mechanism for satisfying time constraints and event response requirements.   Consideration has to be given to the actual time taken to carry out actions.   Any technique for program validation must therefore include an assessment of the performance of the underlying system hardware.

1.3.4 Concurrency and synchronization

In recent years the problems of concurrency and synchronization in multi-tasking software have been studied extensively. Detailed discussions and analysis of the problems can be found in the books listed in the bibliography, a summary of the problems is given in Bennett (1988).

A problem which is fundamental to the use of concurrency is the so called mutual exclusion problem. In simplified terms it is necessary to ensure that tasks which are potentially concurrent and which share a resource of the computer, for example a common memory area, a disc file, or a printer, do not simultaneously attempt to use it. The following scenario gives an example showing how easily mutual exclusion problems can arise.

Two software modules, *bottle_in_count*, and *bottle_out_count* are used to count pulses issued from detectors which observe bottles entering and leaving a processing area. The two modules run as independent tasks. The two tasks operate on the same variable *bottle_count*. Module *bottle_in_count* increments the variable and *bottle_out_count* decrements it. The modules are programmed in a high level language and the relevant program language statements are:

> *bottle_count := bottle_count + 1; (bottle_in_count)*
> *bottle_count := bottle_count - 1; (bottle_out_count)*

At assembler code level the high level instructions become:

> *bottle_in_count*   *bottle_out_count*
> *LD A, (bottle_count)*   *LD A, (bottle_count)*
> *ADD 1*   *SUB 1*
> *LD (bottle_count), A*   *LD (bottle_count), A*

Now if variable *bottle_count* contains the value 10, *bottle_count_in* is running and executes the statement *LD A, (bottle_count)* then as figure 3 shows the *A* register is loaded with the value 10. If the operating system now re-schedules and *bottle_out_count* runs it will also pick up the value 10, subtract one from it and store 9 in *bottle_count* as is shown in figure 3. When execution of *bottle_in_count* resumes its environment will be restored and the *A* register will contain the value 10, one will be added and the value 11 stored in *bottle_count*. Thus the final value of *bottle_count* after adding one to it and subtracting one from it will be 11 instead of the correct value 10.

One solution to this problem is to force the operating system to treat the high level language statements as indivisible operations. A variety of techniques are available but all in principle treat the code statements needing to be protected as <u>critical sections</u> and protect the sections from interfering with each other by using semaphores.

Unfortunately the standard technique of using the critical section approach leads to problems in real-time systems with hard time constraints and alternative solutions have to be used (Bennett, 1988; Faulk and Parnas 1983). A solution which is satisfactory in many engineering applications is not to take any steps to prevent the occasional error. An assessment of the probability of an error being generated and the consequences of such an error should be made before adopting this solution.

A reg	bottle_in_count	count	bottle_out_count	A reg
?	LD A,(bottle_count)	10		
10	*context*	10	LD A,(bottle_count)	10
10	*change forced*	10	SUB 1	9
10	*by operating system*	10	LD (bottle_count),A	9
10	ADD 1	9		9
11	LD (bottle_count),A	9		9
11		11		9

Figure 3 Effect of context change (mutual exclusion).

2. DESIGN STRATEGIES

2.1 Single task

The simplest design strategy is to treat the whole of the software system as a single program written using the techniques applicable to a sequential program. The basic program structure for an embedded real-time system is then:

> {system start-up procedures}
> *repeat*
> {control procedures }
> *until forever;*
> {system shut-down procedures}
> *end.*

If t_c is the time taken to complete one cycle of the control loop (*repeat..until*), and t_e is the time interval for response set by the environment, the time constraint can be expressed as:

Hard constraint $t_{c(max)} < t_{e(max)}$

• Soft constraint $t_{c(average)} < t_{e(average)}$

This approach is recommended for simple systems with a small number of inputs where the response time (t_{ei}) for each input is similar; and the time taken for the various paths through the procedures forming the control loop does not vary greatly from cycle to cycle.

As the number of inputs, actions to be performed, and outputs increases, the range of environment response times, t_{ei}, usually increases. It thus becomes more difficult to satisfy the minimum time requirement if all the actions are carried out every control cycle. One strategy is to partition the inputs and actions according to the required environment response time which gives a system:

$$\{U_1..U_i..U_n\} = \{F_1..F_i..F_n\}\{C_1..C_i..C_n\}$$

where U_i, F_i, C_i, $i=1..n$ represent subsets of the outputs, actions and inputs respectively. The subsets are chosen such that the actions within a specified subset have to be completed within a similar response time. The general program structure becomes:

```
{system start-up procedures}
repeat
if <condition_1 > then {action_1};
if <condition_i > then {action_i};
if <condition_n > then {action_n};
until forever;
{system shut-down procedures}
end.
```

The method used to generate the *<condition>* is implementation dependent. It is also assumed that any sharing of data between actions can be handled by the use of common memory areas.

Applying this strategy to distributed systems implies that the subsets may be implemented on separate processors. To maintain the simplicity of the approach communication between the processors needs to be message based with each processor as one of its actions checking to see if a message has been received. Since there can be no synchronization between the tasks the actually message reception and storage must be handled by hardware or by a separate message processor.

2.2 Two-tasks (foreground-background)

A commonly used design approach is to partition the system into two sections, usually referred to as the foreground and background partitions. The typically division is to place the time dependent actions in the foreground and the time independent actions in the background. Alternatively actions with hard time constraints are placed in the foreground and those with soft constraints in the background. The general rule for forming the partition is that the number of actions placed in the foreground partition should be minimized.

An implicit assumption is that in the system there will be a single task in each partition, thus limiting problems of resource sharing and synchronization to resolving conflicts across the partition boundary.

The foreground - background terminology can give rise to confusion since writers concerned with non-real-time systems refer to interrupt routines, cyclic keyboard input routines, and real-time clock routines as background programs; the same usage can also be found in manuals for real-time BASIC. In the literature on real-time systems the majority usage is that the most time critical routines are said to run in the foreground.

2.3 Multiple task

A natural extension of the foreground - background division is to partition the system into many subsets and treat each subset as a separate task. In implementation terms this is the equivalent of dividing the software into a number of separate programs. For independent subsets the software can be built from a number of independent programs each of which can be implemented using standard sequential processing techniques with the additional requirement that some of the programs will need to be synchronized to the environment.

If the subsets are not independent then in addition to any environmental timing requirements, the various programs will need to communicate with each other. Support for communication between concurrent tasks will be required.

The multi-tasking approach developed during a period when it was assumed that the implementation would be on a single processor with a real-time operating system being used to share resources between the tasks. There is, however, nothing in the approach which requires that a single processor be used. The ideas underlying multi-tasking have led to the development of techniques based on the concept of a virtual machine.

2.4 Virtual Machine

We are all familiar with the idea of a virtual machine. When we program in any high-level language we are programming a computer system which is defined by the language and its interface to a particular operating system: we are not aware of the details of the underlying computer hardware. The virtual machine is defined by the structure of the language.

This concept can be carried further and most real-time software design techniques are based on the idea of the existence of a virtual machine. The software is designed for a virtual machine. Implementation then divides into two stages: code the system for the virtual machine; implement the virtual machine on the real machine.

A virtual real-time machine must support:

Tasks: these are units which perform an action or group of actions and which can be manipulated as an entity. It is assumed that tasks can be created, started, stopped, delayed, deleted i.e. they have the attributes which would normally be associated with a program unit in sequential programming. No assumption is made as to the relationship between the number of tasks and the number of hardware processors; this is left as an implementation decision.

Communication: there must be one or more methods of communication between tasks and between a task and the environment. For example methods of message passing and data sharing may be specified.

Synchronization: methods of synchronization between tasks and between a task and the environment must be provided. This may include explicit provision for the synchronization of a task to a specified standard real-time clock task.

Based on the simple concepts given above a wide range of virtual machine architectures including distributed systems can be generated. An important and interesting approach is the Rex architecture (Baler and Scallon, 1986). This architecture treats application procedures as if they were indivisible individual high level language instructions. That is an application procedure once started is guarantied to run to completion without being interrupted by another application level procedure. In this way data sharing at the application level is simplified. The penalty is that application procedures have to be short. The order in which procedures are run and the interconnection of procedures and data sets is determined by separate program modules.

The virtual machine approach has the merit of delaying implementation issues to a later stage of the design and is strongly recommended as the normally method, even if at the implementation stage it is decided to use a single program technique.

2.4.1 Implementation of virtual machine

There are two basic approaches to implementing a virtual machine:

(i)   map the virtual machine onto a general purpose real-time operating system;

(ii)   build the virtual machine elements in a high level language from a minimal set of primitives provided by a small system kernel.

General purpose operating systems relieve the implementor of many of the chores associated with for example memory management, task scheduling, task communication, interrupt handling. They are, however, restrictive in that they are normally constructed as a monolithic monitor and place severe limitations on the way in which a system can be implemented. For a review of this approach see Gertler and Sedlak (1983).

For simple, non-critical systems (non-critical in the sense that failure will result in inconvenience not serious loss) the combined language-operating systems provided by the so called real-time BASIC languages should be considered. Such systems can also be useful for prototyping.

The development of high level languages which support concurrency has permitted the development of systems with only a small amount of fixed operating system software - referred to as a kernel or nucleus. Additional functions can be·made available as separate modules provided in a library. The user has the choice of developing specific functions for his application or using functions from a standard library.

The latter approach has the advantage of flexibility but the disadvantage of lacking any form of standardization. Suggestions have been made for specific minimum sets of primitives for real-time systems and the IEEE has issued a trial standard giving minimum sets for specific purposes, recommended interfaces to the functions, and bindings to specific languages (IEEE, 1985).

2.5 Design Guide-lines

In real-time software as in general software there is an emphasis on sub-division and the same basic approaches to sub-division can be applied. These are:

* Information hiding;

* Coupling and cohesion and interface minimisation.

Details of the various techniques and their advantages and disadvantages are given in most of the recent texts on Software Engineering and are not covered further here.

For real-time systems some additional rules of guidance are needed as follows:

(i)   Separate actions into groups according to whether the action is:

* time dependent;
* synchronized;
* independent;

and try to minimize the size and number of modules containing time dependent actions.

Divide the time dependent actions into:

* hard constraint;
* soft time constraint;

and try to minimize the size and number of modules with a hard time constraint.

(iii)   Separate actions concerned with the environment from other actions.

The recommended design strategy can be expressed simply as: minimise the part of the system which falls into the category of having a hard time constraint.

It is frequently possible to use properties of the environment to change a hard time constraint into a soft time constraint. Figure 4 shows an example of this and of the separation of the environment actions from the main control software. The control action is assumed to be part of a modulating feedback control loop and to require a specific sampling rate. Normally the control action task would be considered as having a hard time constraint with the following sequence of events having to take place each sample interval:

> *loop*
> *{wait for timing signal}*
> *{read inputs}*
> *{calculate control output}*
> *{transmit control output to plant}*
> *end;*

Using the above sequence the plant inputs will be read at the fixed sample rate. However, there is no guarantee that the plant outputs will be sent out at the required sample interval since the time delay between reading the plant inputs and sending back the outputs will vary with the computational time. An improvement is obtained by changing the sequence of actions to

```
loop
{wait for timing signal}
{output control variable m(k)}
{read input variable c(k+1)}
{calculate control variable m(k+1)}
end;
```

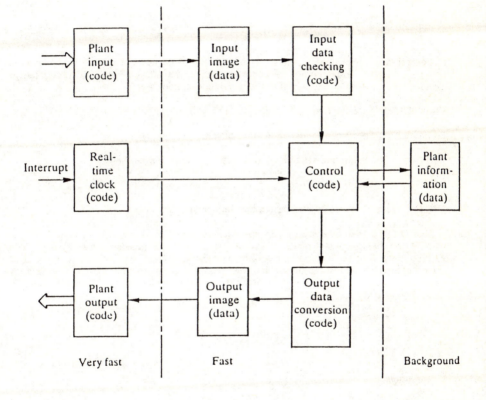

Figure 4   Separation of environment actions from main control actions.

With this sequence the variation in the time taken to calculate the control variable does not affect the rate at which it is transmitted to the plant. The fixed delay of one sampling interval can be taken into account in designing the actual control algorithm.

Changing the order of the actions de-couples the control actions from the input-output actions thus separating the environment from the internal activities. The plant input activity creates an <u>input image</u>: the output activity transmits the <u>output image</u> to the plant. In order to maintain synchronization with the environment the plant input and output task must run with a hard time constraint. The time constraint on the control task usually can be relaxed: either to the requirement that it completes at sometime within the sample interval or, since in many applications the occasional missed sample will have little effect on the plant, to a requirement that on average it completes within the sample interval.

3. DESIGN TECHNIQUES AND TOOLS

3.1 Introduction

Several design techniques and tools have been developed specifically for real-time systems, ranging from purely specification techniques through to full development systems with or without construction tools. Some examples are:

RSL/REVS   Specification and simulation tools (Alford, 1977)

PAISLey   Specification and simulation tools (Zave, 1982)

DARTS   Design and analysis of real-time systems (Gomaa 1984)

MASCOT   Design, construction, operation and test tools

SDRTS Structured development, design and implementation system.

The DARTS method is similar in approach to the SDRTS method. The RSL/REVS method has being widely reported in general Software Engineering text books. The PAISLey technique is an important development in methods of specifying software and in software development techniques; it relies heavily on formal methods and as such is outside the scope of this Chapter. A good overview of the method is in Zave (1982).

4. MASCOT

4.1 Outline

The first version of Mascot was developed by Jackson and Simpson during the period 1971-75 (Jackson and Simpson 1975). The official definition of Mascot 1 was published in 1978 and a revised version - Mascot 2 - was issued in 1983. Between 1983 and 1987 extensive changes to the technique were made and the official standard for Mascot 3 was published in 1987. The discussion which follows relates to Mascot 3.

The official handbook states that:

"Mascot is a Modular Approach to Software Construction Operation and Test which
    incorporates:
    a means of design representation

a method of deriving the design
a way of constructing software so that it is consistent with the design
a means of executing the constructed software so that the design structure remains visible at run time
facilities for testing the software in terms of the design structure."

In Mascot software is represented as a system - a combination of

(i)   a set of concurrent functions and
(ii)   the flow of data between such functions.

The functions are referred to as components and the software system is represented as a set of interconnected but independent components which make no direct reference to each other.   Each component has specific, user defined, characteristics which determine how it can be connected to other components.

Components are created from templates, that is patterns used to define the structure of the component.   Two classes of templates are fundamental to Mascot (i) activity (ii) intercommunication data area - IDA.

An activity template is used to create one or more activity components each of which is a single sequential program thread that can be independently scheduled.   It is assumed that at the implementation stage each activity will be mapped onto a software task.   Such a task may run on its own processor or be scheduled by a run-time system (usually referred to as the MASCOT kernel) to run on a processor shared with other activities. The activities communicate through IDAs.   The IDA provides the necessary synchronization and mutual exclusion facilities.

An IDA is a passive element with the sole purpose of servicing the data communication needs of activity components.   It can contain its own private data areas. It provides procedures which activities use for the transfer of data.   Within an IDA, and only within an IDA, the designer has access to low level synchronization procedures. This feature allows a Mascot designer to use any technique appropriate to the problem. He is not limited to using high-level operations such as monitors, message-passing, or rendezvous, provided by the implementation language.

A structure containing activity components connected by means of one or more IDAs is referred to as a network.

4.2 Simple Example

Mascot supports three forms of IDA: a generalized IDA; a channel; and a pool.   The graphical symbols for each are shown in figure 5 .

The channel and the pool will be familiar to users of Mascot-2 - their behaviour is defined as follows:

channel   supports communication between producers and consumers.   It can contain one or more items of information. Writing to a channel adds an item without changing items already in it.   The read operations is destructive in that it removes an item from the channel.   A channel can become empty and also, because its capacity is finite, it can become full.

pool   is typically used to represent a table or dictionary which activities periodically consult or update.   The write operation on a pool is destructive and the read operation is non-destructive.

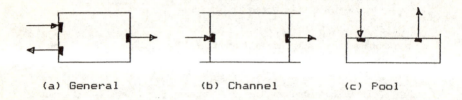

(a) General   (b) Channel   (c) Pool

Figure 5 Graphical symbols for IDAs

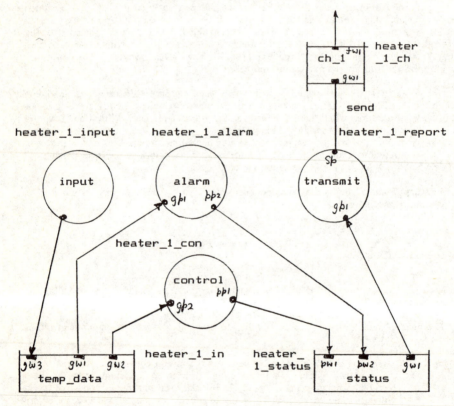

Figure 6 Example of MASCOT ACP diagram

Mascot can be used at a simple level to provide a virtual machine supporting activities, pools and channels. A design is constructed in the form of an activity, pool, and channel network - an ACP diagram - as is shown in figure 6.

The diagram represents part of a system for the control a plant.   The activity *heater_1_input* gets data from a plant interface. The data is held in a pool *heater_1_in* from where it is read by activity *heater_1_alarm* and *heater_1_con*.   The required output to the plant and the alarm status are held in a pool *heater_1_status*. An activity *heater_1_report* gets data from the pool holding status information and sends it via a channel *heater_1_ch* to some other activity (not shown).   Also not shown are the activities required to pass the data to the plant control. This ACP differs from a Mascot 2 ACP since the components now contain ports and windows (shown as filled in circles and rectangles).

Once the ACP diagram has been produced design of the templates for the individual components can proceed. Many component templates will be reusable and hence only application specific ones will need to be designed.   Instances of the component are created when the network is constructed by translating the ACP diagram to textual form and entering it into the MASCOT database.

At this level a design in Mascot may be represented either in graphical (ACP diagram) or textual form.   Both forms are equivalent and may be derived from each other.   The textual form stored in the database can be progressively updated as the design proceeds.

A major limitation of MASCOT 2 was the absence of facilities for representing hierarchical structures.   A system was represented as a two dimensional data flow network of alternate data processing and data communication elements.   Higher level elements could be utilised in developing the network but they were not retained in the database and were not recognized as design entities.   A large network could be partitioned into several arbitrary subsidiary networks the connection between adjacent networks was through the sharing of one or more communication elements.   The subsidiary network thus formed was treated as a unit for control purposes at run-time.

It also supports at the design stage means of specifying use of direct data visibility, often a necessity in real-time software. A typical example is a module providing direct feedback control subjected to a hard time constraint. If a module of this type requires access to external data, for example in order to update controller parameters, it must have guarantied access at all times and must not be kept waiting because another module is accessing the data. There are a variety of solutions to this problem but one is to allow the module to access the data directly without using the standard access procedures.

Mascot provides the designer with a means of providing direct access through a construct called an access interface.

4.3 Run-time features

On any given system software can normally be divided into application specific modules and modules which form part of the environment in which the software is running. Mascot, while not specifying any mandatory requirements for run-time support, assumes that the environment will provide a variety of support functions.

Any support functions needed must be specified the functions in a context interface module.   The functions can be unqualified in which case they are available to all classes of template modules or they may be qualified in which case they are restricted to those modules listed in the qualification list.

At the design level Mascot makes no assumptions about how the activities are to be run and intercommunication organised. For the earlier versions of Mascot applications typically ran on a single processor and were programmed in a language which did not support concurrency: this is no longer true.

If the language being used in the implementation supports concurrency then the implementor should consider mapping activities onto the appropriate language feature. Mascot imposes one restriction: activities should not be created dynamically, the system network (activities, IDAs and servers) must remain invariant at run-time. The implementor must document how the language features have been used to support the Mascot virtual machine.

Rather than impose specific schemes for synchronization, device handling, interrupts, process scheduling and priorities, a model scheme is suggested. Adoption of this scheme is optional. An analysis of the scheme and comparison with alternatives is given in Sears and Middleditch (1985). Bugden (1985) has described how it can be implemented in Modula 2.

4.4 Summary

Mascot 3 provides and excellent design methodology. It is sufficiently rich in concepts to provide flexibility for design while still providing constraints which assist in creating safe and reliable software. The hierarchical structure introduced in Mascot 3 has overcome the limitations of the previous versions.

It provides a good mix of graphical and textual notations. The incremental approach to creating software modules is of great assistance in separating design from implementation. Even if the full project support system is not used the design techniques and notations provide a powerful free standing design technique which coupled with the use of modern languages such as Modula 2 or Ada for implementation can provide an effective means of creating real-time software.

The project support environment - the Mascot database -provides a methodology for building the software. It also provides the designer with some simple checks on consistency. However the facilities are limited and at least one attempt is being made to extend the support given to the designer (Moorhouse, 1986).

A weakness of the system, but one which is true of most design methodologies, is that it has no facilities for representing or assisting with the design or implementation of error detection or error recovery.

Mascot provides very little guidance for the developer on the difficult issues of mapping the virtual model onto a real machine or machines. The formal methodology assumes that a design can be expanded down to the level of activities containing root and subroot modules without any consideration of how the virtual machine will be implemented. This is unrealistic since any expansion beyond subsystem modules is likely to impinge on implementation issues. In practice subsystem expansion will be based, at least implicitly, on assumptions about the method of implementation and will be revised in the light of implementation changes.

The next method considered attempts to draw a sharp boundary between two phases of design activity: the first phase where implementation considerations are rigourously excluded and the second in which implementation considerations are explicitly considered.

5. SDRTS: STRUCTURED DEVELOPMENT FOR REAL-TIME SYSTEMS

5.1 Introduction

The ideas underlying this method are similar to those utilised in MASCOT. It is assumed that the starting point of any design is to build a software model represents the system requirements in terms of the abstract entities supported by a virtual machine. This model is called the essential model. The second stage of the design is to derive from the essential model an implementation model and in doing so determine on a rational basis some implementation decisions. For example the number of processors required, the memory requirements, the mapping of activities to processors and tasks within a processor.

The technique, as its name implies, is a significant extension of the structured design methodology. The notation used is an extension of standard flow diagram notation.

5.2 Modelling notation and techniques

As in standard flow diagrams, bubbles represent transformations and directed lines represent data flows. However, the notation distinguishes between different types of data flow as is shown in figure 7.

Continuous data may be in analogue or digital form. It is represented by the double arrow head: *current_pH* and *valve_control* are continuous data flows. Data transformations of continuous data flows for example *change_pH* are assumed to operate continuously. If the data transformation is implemented digitally then the frequency with which it runs must be sufficient to approximate to continuous operation.

Discrete data may be analogue or digital although it will normally be digital. It is indicated by a single arrow head as shown on *pH_demand*. Data transformations operating on discrete data flows are assumed to be triggered by the arrival of a unit of data - a transaction.

Event data flows are shown by means of dashed lines. Events are data flows which do not have numeric content but simply indicate that an event has occurred or provide a signal or command. Thus *pH_at_desired_value* is a signal, and *start* and *stop* are commands. Event flows are processed by control transformations which are indicated by a bubble with a dashed outline. Control transformations can issue special event flows known as prompts and triggers. The enable/disable event flows are prompts. They are used to enable and disable data transformations. Triggers can be used to force a data transformation to run in response to a specific event or combination of events.

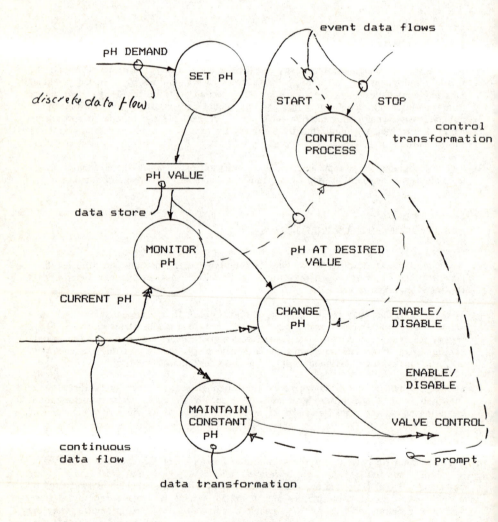

Figure 7 Example of design using SDRTS notation

5.2.1 Modeling Conventions

The modelling method assumes that the designer will adopted certain conventions which can be summarised as:

Control transformations:	inputs –	event flows
		prompts
	outputs –	event flows
		prompts
Data transformations:	inputs –	event flows
		prompts
	outputs –	data flows
		event flows

Only control transformations may generate prompts.

Data transformations may, by generating event flows, control events outside the software system but only control transformations may control (prompt) activities within the software system.

5.3 Essential modelling

The essential model is created by following the procedure outlined in figure 8. The method emphasises the separation of the environment from the internal structure. The normal first stage in building the abstract model is to produce a context schema and event list to describe the environmental model.

An example of a context schema is given in figure 9. The terminal units shown in rectangular boxes e.g. *plant_A10_input_interface* can be treated as virtual devices at this level rather than the actual device.

The modelling of an actual transducer to a virtual transducer is illustrated in figure 10. This convention fits in well with the idea of creating an 'image' of the external environment upon which the system software operates. The 'image' has known characteristics and isolates the system from interface complications and vagaries.

The transformation *plant_control* is expanded into the behavioural model either by means of transformation schema or data schema: the choice depends on the nature of the problem. A possible transformation schema is shown in figure 11.

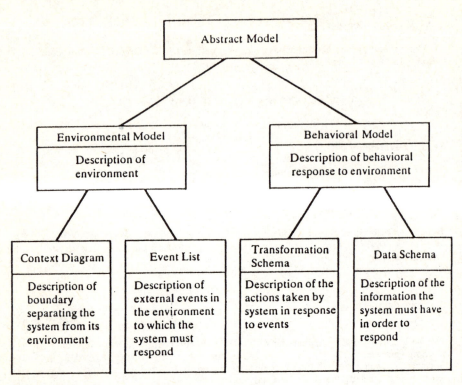

Figure 8   Outline of the abstract modelling schema

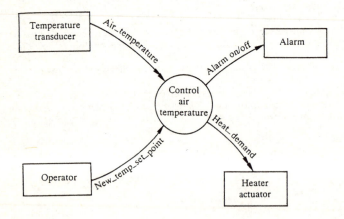

Figure 9 Example of a context schema

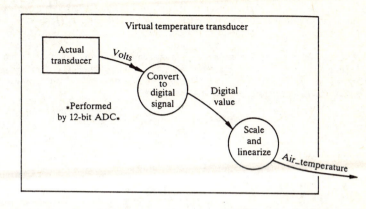

Figure 10 Modelling of interface transducer

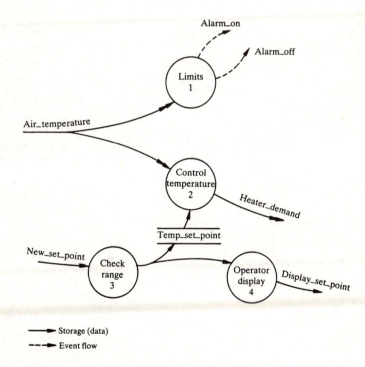

Figure 11 Transformation schema

This is not a complete system as for example there is no data flow to the data store for *limit_values*. The trigger prompt is used to indicate that *control_temperature* will act at intervals determined by the presence of a signal on trigger. Thus by firing the trigger from a clock the transformation could be specified to run at predetermined intervals.

The design procedure encourages a hierarchical approach as the elaboration of each transformation can be considered as creating a subsystem. The designer has to use her judgment to decide when to stop further elaboration. When this point is reached each transformation must be specified in some manner.

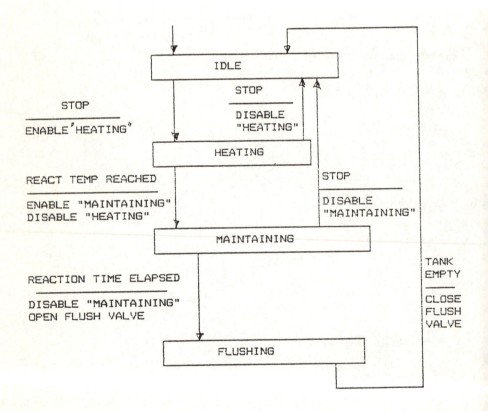

Figure 12 Example of a state transition diagram

For specifying control transformations Ward and Mellor suggest the use of state transition diagrams and action tables. Their approach is based on the work of Hopcroft and Ullman Hopcroft and Ullman, 1979). An example of the notation they use is given in figure 12. The state transition technique is a powerful method and Ward and Mellor are only able to give limited coverage; more detailed explanations of the method can be found in Fairley (1985).

Ward and Mellor (1986) discuss a number of methods of specifying data transformations including the standard procedural techniques of program design languages, pseudo-code and structured English. They also discuss and give examples of a non-procedural method based on the use of precondition-postcondition statements (Heniger, 1980). If the goal of producing an essential model devoid of implementation constraints is to be achieved it would seem important that specifications at this stage are expressed non-procedurally. The choice of methodology is open: any of the rapidly developing formal techniques can be used.

5.4 Checking the essential model

Ward and Mellor recommend checking the transformation schema of the model in two ways. The first is to use the rules for data flow to check for consistency. This is the equivalent of checking the syntax of a program and can be done by hand or, given the advances in graphics processing capabilities in recent years, it is now feasible to construct a graphics compiler to perform the necessary checks.

The second level of checking is to determine whether the model can be executed - can it in some sense generate outputs from a given set of inputs. The approach suggested by Ward and Mellor is based on ideas derived work on Petri nets.

The method is based on the use of tokens. The presence of a token indicates that a data flow has a value: the absence of a token indicates that it has no value. The execution is carried out by tracking the propagation of tokens through the system from some given starting point. It is assumed that provided the requisite tokens are present a data transformation will produce the appropriate output tokens. For control transformations the production of tokens is determined by execution of the state transition table specifying the transformation.

The procedure can be carried out by hand, however, it is preferable to have software capable of carrying out the procedures.

At this stage the abstract modelling is complete and we are ready to move to the implementation stage.

5.5 Implementation

The first stage of implementation is to map the essential model onto the implementation model. This is the process of examining possible implementation technologies and choosing the ones which introduce minimum distortion into the mapping. Ward and Mellor illustrate what they mean by minimum distortion using as an example a satellite attitude modification system. On a modern satellite the system would be implemented as a single task on a on board microprocessor and the implementation model would thus involve no distortion as the data transformation 'modify attitude' would map to a task.

In the 1960s it would not have been possible to have an on board computer and the attitude would have been controlled from a ground station. The implementation model would thus have needed to split 'modify attitude' into a device to transmit signals to the ground placed on board the satellite and a processor on the ground. There would not be a simple one-to-one mapping between the essential and implementation models and hence distortion is introduced.

Ward and Mellor suggest following a top down approach in constructing the implementation model.

* allocate units of the essential model to processors;
* allocate units to tasks within a single processor;
* allocate units to modules within a single task.

A processor is defined as a person or machine that can carry out instructions and store data. Machines may be anything from simple analogue or digital circuits through to supercomputers.

A task is a unit of code which can be started, stopped, delayed, interrupted, suspended and resumed by the system software running on a single processor. The implementation model relevant to a given processor will consist of a network of tasks (the network may be a single task). Tasks are assumed to be concurrent.

A module is a segment of code which is treated as a unit by a task. A task is assumed to activate modules in a mutual exclusive manner.

To answer some of these questions it is necessary to elaborate the some of the time constraints which should have been given in the specification.

6. TIMING CONSIDERATIONS

6.1 Example 1

As an example consider a system used to provide feedback control on a plant. The actions to be performed are:

	Action	Cycle Time	Compute Time (Max)
P1	read plant inputs	t_p	t_1
P2	check alarm status	t_p	t_2
P3	compute control outputs	t_p	t_3
P4	output to actuators	t_p	t_4
P5	update operator display	t_d	t_5
P6	check operator switches	t_o	t_6
P7	output log information	t_l	t_7

The cycle time is the repeat time set by the environment: t_p is a hard constraint i.e. the plant inputs must be read, the control value calculated and sent to the actuators every t_p seconds. The values t_d, t_o and t_l represent soft constraints in that on average the actions such be performed at those intervals.

Using a single task method all the actions P1 to P7 would have to be carried out every t_p seconds which gives a condition for correct operation

$$t_p > t_1 + t_2 + t_3 + t_4 + t_5 + t_6 + t_7$$

Using a foreground - background approach actions P1, P2, P3, P4 would form the foreground partition with actions P5, P6 and P7 forming a background partition. Necessary conditions for correct operation are:

1. $$t_p > t_1 + t_2 + t_3 + t_4 + \frac{(\overline{t_5} + \overline{t_6} + \overline{t_7})}{\min\{t_d, t_o, t_l\}}$$

2. $$\min\{t_d, t_o, t_l\} > \overline{(t_5 + t_6 + t_7)}$$

where $\overline{t}$ represents the average computation time not the

maximum.

The above assumes that the system is run on one processor and that actions P5, P6 and P7 are run as a single task in the background and can be pre-empted by the foreground task.

If a multi-task approach is used then actions P1, P2, P3 and P4, because of their similar time scales, could be grouped as a single control task. With actions P5, P6, and P7 treated as separate tasks. Necessary conditions for correct operation then become:

1. $$t_p > t_1 + t_2 + t_3 + t_4 + \frac{\overline{t_5}}{t_d} + \frac{\overline{t_6}}{t_o} + \frac{\overline{t_7}}{t_l}$$

2. $$t_d > t_c \cdot t_d / t_p + \overline{t_5} + \overline{t_6} \frac{t_d}{t_o} + \overline{t_7} \frac{t_d}{t_l}$$

3. $$t_o > t_c \cdot t_o / t_p + \overline{t_5} \frac{t_o}{t_d} + \overline{t_6} + \overline{t_7} \frac{t_o}{t_l}$$

4. $$t_l > t_c \cdot t_l / t_p + \overline{t_5} \frac{t_l}{t_d} + \overline{t_6} \frac{t_l}{t_o} + \overline{t_7}$$

where t_c is the average time taken to carry out actions P1, P2, P3, and P4.

By making use of the self correcting properties of modulating feedback control it is possible to partition the control actions in a different way as is illustrated in figure 4. This partitioning is based on the assumption that if occasionally a newly computed control value is not available the disturbance to the system will be minor. On this assumption the hard time constraint is applied to the input, output and alarm actions, the control calculation itself is allowed to fall into a high priority 'on average' category.

6.2 Example of cyclic tasks

Three tasks A,B and C are required to run at 20mS, 40mS and 80mS intervals.

(Corresponding to 1 tick, 2 ticks and 4 ticks, if the clock interrupt rate is set at 20 mS). If the task priority order is set as A, B and C with A as the highest priority then the processing will proceed as shown in figure 13a with the result that the tasks will be run at constant intervals. It should be noted that using a single CPU it is not possible to have all the tasks starting in synchronism with the clock tick. All but one of the tasks will be delayed relative to the clock tick, but the interval between successive invocations of the task will be constant.

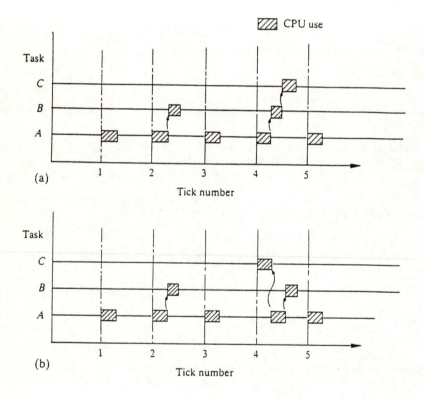

Figure 13   Task activation diagram (a) priority A,B, C (b) priority C,A,B

If the priority order is now re-arranged so that it is C,A and B then the activation diagram is as shown in figure 13b and every fourth tick of the clock there will be a delay in the timing of tasks A and B. In practice there is unlikely to be any justification for choosing a priority order C,A and B rather than A,B and C. Usually the task with the highest repetition rate will have the most stringent timing requirements and hence will be assigned the highest priority.

A further problem which can arise is that a clock level task may require a longer time than the interval between clock interrupts to complete its processing (note that for overall satisfactory operation of the system such a task cannot run at a high repetition rate).

6.3 Example of timing of cyclic tasks

Assume that in the previous example task C takes 25 mS to complete, task A takes 1 mS and task B takes 6 mS. Figure 14 shows the activity diagram if task C is allowed to run to completion and task will be delayed by 11 mS every fourth invocation. It is ;normal to divide the cyclic tasks intro high priority tasks that are guarantied to complete within the clock interval and lower priority tasks that can be interrupted by the next clock tick.

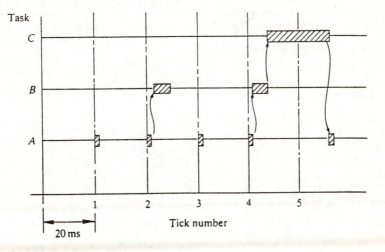

Figure 14 Task activation diagram

The real-time clock handler, which acts as the dispatcher for the system and controls the activation of the clock level tasks, must be designed carefully as it is run at frequent intervals.· Particular attention has to be paid to the method of selecting the tasks to be run at each clock interval. If a check of all tasks were to be carried out then ·the overheads involved could become significant. A method of selecting the appropriate task is given in Bennett (1988, pp. 206-207).

The virtual machine approach to design enables processes to be coded and the computation time for a process to be estimated by running on a target processor, running on a simulator, or by inspection of the code. Of course some assumption about the overheads involved in communication and task switching will have to be made as the details of these are not available until decisions on the system structure are finalised. Various performance modelling techniques can also be used to evaluate and refine designs (Vittins and Signer 1986).

7. CONCLUSIONS

We have reviewed some of the issues relating to the development and design of real-time software and briefly outlined two design methodologies. With the increased use of microprocessors in equipment there is a growing need for software engineers and other engineers to be familiar with techniques for developing real-time software.

Implementors are moving away from reliance on monolithic, general purpose, operating systems and are using minimum operating system kernels. The additional operating system features required are then built using a high level language for a particular application or group of applications. There is also an increasing use of multi-processor systems with a consequent increased concern with communications, distributed data-bases and distributed operating systems.

8. REFERENCES AND BIBLIOGRAPHY

Alford, M.W., 'A requirements engineering methodology', *IEEE Trans. Software Engineering*, *SE-3*, 180-193 (1977).

Allworth, S.T., Zobel, R.N., *Introduction to Real-time Software Design*, Macmillan, London (1987)

Baler, T.P., Scallon, G.M., "An architecture for real-time software systems", *IEEE Software Magazine*, 3, 50-58. (1986).

Bennett, S., *Real-time Computer Control: an Introduction*, Prentice Hall, Englewood Cliffs NJ, (1988).

Bennett. S., Linkens, D.A., (editors) *Real-Time Computer Control*, Peter Peregrinus, Stevenage, (1984)

Budgen, D., 'Combining MASCOT with Modula-2 to aid the engineering of real-time systems', *Software - Practice and Experience*, 15, 767-793 (1985).

Fairley, R., *Software Engineering*, McGraw-Hill, New York, (1985).

Faulk, S.R., Parnas, D.L., "On the uses of synchronization in hard-real-time systems", *Proc.of the Real-time Systems Symposium, IEEE*, Arlington, VA. 6-8 Dec. 101-109 (1983)

Gertler, J., Sedlak, J., "Software for process control - a survey" in Glass, R.L. (editor), *Real-time Software*, Prentice Hall, Englewood Cliffs NJ, 13-44, (1983).

Glass, R.L., *Real-Time Software*, Prentice-Hall, Englewood Cliffs NJ, (1983)

Gomaa, H., 'A software design method for real-time systems' *Comm ACM*, 27 938-949 (1984).

Guth, R., *Computer Systems for Process Control*, Plenum Press, New York, (1986)

Hatley, D.J., Pirbhai, I.A., *Strategies for Real-time System Specification*, Dorset House, New York, 1987

Heniger, K.L., 'Specifying software requirements for complex systems: new techniques and their application', *IEEE Trans. Software Engineering, SE-6* 3-13 (1980)

Hopcroft, J.E., Ullman, J.D., *Introduction to Automata Theory, Languages, and Computation*, Addison-Wesley, Reading, 16-45, (1979).

IEEE, *IEEE Trial-use Standard Specifications for Microprocessor Operating Systems Interfaces*, Wiley, New York, (1985).

Jovic, F., *Process Control Systems: Principles of Design and Operation*, Kogan Page, London, (1986)

Lawrence, P.W., Mauch, K., *Real-time Microcomputer Design*, McGraw-Hill, New York, (1987)

Mellichamp, D., (editor) *Real-Time Computing with Applications to Data Acquisition and Control*, Van Nostrand, New York, (1983)

Moorhouse, T.J., "MDSE Concepts", Ferranti Computer Systems, Alvey Project Document MDSE/GEN/TN/F3.4 July (1986).

Pyle, I.C., "Methods for the design of control software", in *Software for Computer Control. Proc. Second IFAC/IFIP Symposium on Software for Computer Control*, Prague 1979, Pergammon, Oxford, (1979).

Sears, K.H., Middleditch, A.E., 'Software concurrency in real-time control systems: a software nucleus', *Software - Practice and Experience*, 15, 739-759 (1985).

Vitins, M., Signer, K., "Performance modelling of control systems", in R. Guth (editor), *Computer Systems for Process Control*, Plenum Press, New York, 141-167, (1986).

Ward, P.T., Mellor, S.J., *Structured Development for Real-Time Systems*, 3 vols. Yourdon Press, New York, (1986)

Wirth, N., "Towards a discipline of real-time programming", *Comm. ACM*, 22 577-583 (1977).

Zave, P. 'An operational approach to specification for embedded systems', *IEEE Trans. Software Engineering, SE-8* 250–269 (1982).

Chapter 9

Real-time software using MASCOT

N. Mort

1. INTRODUCTION TO MASCOT

The name 'MASCOT' is an acronym for Modular Approach to Software Construction, Operation and Test and, as acronyms go, this one is accurate and useful without succeeding in telling quite the whole story. It provides a number of significant keywords on which a description can be based. This introductory section attempts to lay the foundation for the details of MASCOT which are described in later sections.

First, it is important to establish the types of computer systems that MASCOT users are likely to be dealing with, namely large scale and complex systems. They are large scale in the sense that the software can only be implemented in an acceptable time scale by employing a large team of programmers. They are complex in the sense that they have to perform a large number of parallel tasks which interact with each other in a manner which is subject to real time constraints. Such systems are commonly required for military applications and reflect the origins of MASCOT itself (Jackson and Simpson (1975). In general terms, the purpose of these large, complex computer systems is to assist human operators to make the most effective use of equipment whose function is to monitor and control aspects of the environment through a collection of sensors and actuators.

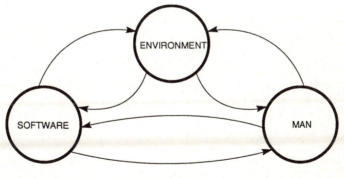

FIGURE 1.1

Figure 1.1 illustrates the interactions involved. The operators and the software exchange information and collaborate with each other in order to achieve, via the sensors and actuators, a more effective interaction with the environment.

The Air Traffic Control System illustrated schematically in Figure 1.2 provides a more concrete example.

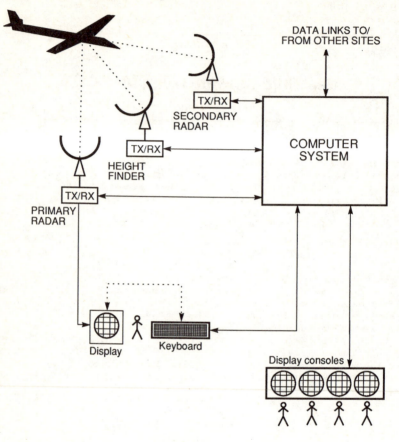

FIGURE 1.2

Three types of radar independently scan the air space (the environment) and each of these feeds information to the computer system. Mutually asynchronous processes within the software are needed to handle these independent sources of data. Further processes are responsible for sending out data to effect controlled manipulation of the height finder and secondary radars. In parallel, operators, observing the results of the primary radar scan directly, are feeding instructions into the computer system via on-line terminals. This imples other software pro-cesses and yet more processes organise and transmit the information to be displayed to the operations controllers at their consoles, react to instructions from the controllers and relay information over data links to other sites. Thus it can be seen that the software is made up of a potentially large number of continuous tasks which have to be performed independently and in parallel but which nevertheless interact with each other through the exchange of data. In this particular instance, changes in the environment take place through the agency of actuators (the aircraft pilots) which are outside the system. Their actions are, however, based on information derived from the system via radio communication between controllers and aircraft. If the task of creating software to satisfy a requirement of this kind is to be carried through successfully and efficiently, a high degree of systematic organisation is needed and MASCOT is one way of satisfying this requirement.

2. THE MASCOT KERNEL

MASCOT is, first and foremost, a consistent methodology to be applied to the complete lifecycle of a software project from the initial understanding and clarification of the requirement through to the maintenance and extension of the operational system. Central to this whole philosophy is the computer environment in which MASCOT application programs function. This is closely defined in the Official Definition (MASCOT Handbook (1980) and is provided by a compact piece of supervisory software known as the MASCOT Kernel.

It is useful at this stage to give a brief summary of the facilities which a MASCOT Kernel is expected to provide. These fall under four headings:

Scheduling

The kernel must determine, from time to time, which of the set of parallel tasks consituting a MASCOT system is to receive the next available quantum of processing power. It must be aware of the passage of real time and it must provide for the execution, on behalf of the tasks, of a set of synchronising primitive operations.

Interrupt Handling

However this is implemented, and there are several options available, it is the kernel which must perform the immediate action in response to a hardware interrupt.

Subsystem Control

The kernel provides the operator of a MASCOT system with the means of introducing to and removing (temporarily or permanently) from the attention of the scheduling algorithm, groups of related tasks known as subsystems.

Monitoring

A comprehensive and flexible set of monitoring facilities are defined to aid testing and optimisation.

3. MODULARITY IN MASCOT

To understand MASCOT it is convenient to start with the MASCOT concept of modularity. Experience has shown that it is essential to have modularity at several stages in the design and implementation of software. However, the reasons for needing modularity differ between these stages and this has led to inconsistent forms of modularity being adopted at the various stages. The need for modularity at the design stage arises out of the limitations of the human brain. It is possible to cope with the intricacies of a large system successfully only by breaking it down into functional units. Even then the task is a difficult one unless there is a suitable 'language' in which to express the developing design. Conventionally, the process of functional decomposition is continued until a a point is reached where the modules are considered small enough to be handed out for coding by individual programmers.

At the software construction phase (compiling, loading etc.) modularity arises in a different way and is associated to some extent with language design. The desirability of building programs from independently compiled modules has been accepted from the early days of programming. It was built into FORTRAN from the beginning with each subroutine being independently compilable. However, the criterion for writing a subroutine at that stage was to accommodate code used several times. Only subsequently has the attraction of the 'structural' subroutine been realised and the relationship between the functional design modules referred to above and these 'modules of compilation' has not really been formalised. At run time a third type of modularity appears in the form of parallel co-operating processes. Each multiprogrammed thread is a 'module of activitity' identifiable as an independent asynchronous process, e.g. now running; now held up awaiting its share of the CPU; now held up waiting for some other parallel thread to provide it with input data to consume its output data; now waiting for an external interrupt. This form of modularity is forced on the user by the nature of the system being addressed.

4.   MASCOT ACTIVITIES AND IDAs

It is a cornerstone of the MASCOT approach, from which much else follows, that the three disparate forms of modularity referred to in the previous section should be unified into a single form. Furthermore, it is the run time 'module of activity' which is the basis of the unification and for the remainder of this chapter, it will be referred to simply as an 'activity'. Thus, the MASCOT formalism for expressing the structure of complex, real-time systems is based on the activity as, successively, a unit of design, a unit of coding, a unit of construction and as an independent, parallel thread in an operational system.

If asynchronous activities are to co-operate in order to form an integrated system, they must be able to communicate with each order through common data areas. The control of access to these data areas is of central importance in ensuring the integrity of the system. In the MASCOT approach, an individual activity's awareness of the rest of the system arises solely from its access to data areas which it shares with a limited number of other activities. Each activity is thus encapsulated in a run-time environment which extends no further than is necessary for it to perform its allotted task. An outstanding advantage of this arrangement is that it makes it relatively easy to execute an activity in a test environment which it cannot distinguish from that in which it will ultimately operate.

Having explored the controlled, co-operative use of memory space, there remains the question of the relative timing of activities which is of equal importance. A MASCOT system is expected to respond appropriately, and within an acceptable time scale, to interrupts which signal the occurrence of significant events in its enviornment. The mechanism by which MASCOT achieves this involves the use of a second type of building block called an intercommunication Data Area or IDA. These contain not only the shared data areas required for inter-activity communication but also the 'mechanisms' for obtaining access to these areas. Any complications involved with synchronisation, etc. are therefore removed from the activities which, therefore, can be coded as normal sequential processes.

In conclusion, it will be noted that the description of MASCOT given in this section has been presented without recourse to any particular programming language. Historically, the MOD origins of MASCOT led to most early implementations being produced in CORAL 66. More recently, it has been suggested that a Modula-2/MASCOT combination is a sound basis for the production of application software for real-time systems (Budgen (1985)). A further possibility is examined in the next chapter when the possibilities of an Ada implementation are discussed.

5. MASCOT NETWORK (ACP) DIAGRAMS

Many of the problems of software development and production, which were discussed in the previous section, reduce to problems of communication. The MASCOT concept of a unified form of modularity, based on the run-time units of activity, opens up the prospect of a unified notation. What is required is a language independent notation which is capable of illustrating unambiguously the access rights of constituent activities to shared data areas. These requirements are met by a network diagram the nodes of which are either activities or data areas, and in which the interconnecting lines represent the valid paths for data flow. The standard form of MASCOT network diagram is derived from the Phillips diagram (Phillips (1967)) and differs in the sense that it uses a relatively few number of symbols. The run-time unit of activity which is the MASCOT module is known as an ACTIVITY and is represented on the diagram by a circle. At the design stage each activity is distinguished by a name which normally indicates its function (see Figure 5.1).

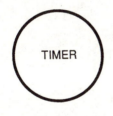 

FIGURE 5.1

The activities communicate with each other and with the outside world through shared data areas called IDAs. It has been found useful to distinguish two classes of IDA which are involved in different modes of data transfer. The first of these is the CHANNEL which, as its name suggests, carries a flow of data. It has two distinct interfaces with activities. One is an input interface at which 'producer' activities enter data and the other an output interface at which 'consumer' activities extract data. The storage of data within a channel is inherently transient; a piece of data read from a channel is removed and cannot be read twice. Normally, the data storage will be organised as a pipe on a first-in-first-out basis. The symbol for a channel is:

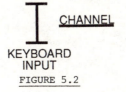

FIGURE 5.2

and, as in the case of the activity, it is usual to append a distinguishing name.

The second type of IDA is called a POOL and has a single interface which
may be used for either input or output by the associated activities. A
pool holds relatively permanent information such as data that might be
held in a table. Reading data from a pool does not remove it from the
pool; there is no concept of data consumption here. The network diagram
symbol is:

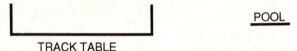

POOL

TRACK TABLE

FIGURE 5.3

Activities and IDAS are joined by lines with arrows indicating the direction of data flow. For example, a simple instance of the common
'producer-consumer' situation could be represented as shown in Figure 5.4
below.

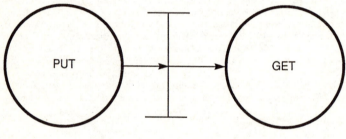

PUT

GET

FIGURE 5.4

These interconnecting lines must not, of course, be confused with the
corresponding lines in a flow chart which represent the movement of
control. Control in a network diagram is hidden; it can be thought of
as existing in another dimension within each activity. However, there
is one aspect of control which sometimes needs to be shown and this is
the point of effect in the system of an external interrupt. Also, it
may be necessary to include a hardware device on the diagram. An interrupt handler reading characters from a tape and passing them into a
channel could be depicted as shown in Figure 5.5:

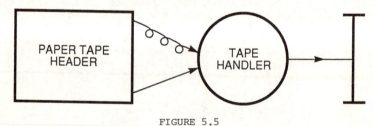

PAPER TAPE
HEADER

TAPE
HANDLER

FIGURE 5.5

MASCOT network diagrams are known as Activity Channel Pool diagrams or, more concisely, as ACP diagrams. The following simple example to illustrate the idea of these diagrams assumes that a computer is in use to control the rate of a chemical reaction. Control is exercised by actuating a valve and the rate of reaction is monitored by a temperature sensor. In addition, there is a keyboard from which an operator may issue commands to increase or decrease the rate and a printer to provide a summary of performance on request. One solution would be to create a control activity, a keyboard activity and a printer activity leading to the ACP diagram shown below in Figure 5.6:

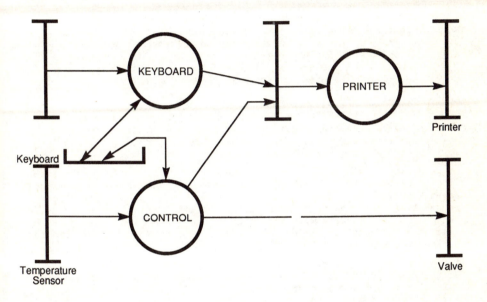

FIGURE 5.6

6. ACTIVITY SCHEDULING AND SUBSYSTEMS

An activity in MASCOT is in the unit of scheduling. Contained within the MASCOT Kernel software is a dispatcher algorithm which determines, from time to time, which of the applications activities is to run next. The number of actual hardware processors available is immaterial to the MASCOT philosophy as an activity, which is ready to run, can be allocated to any available processor. This is important in view of the interest in distributed, multi-processor systems.

The period of CPU activity which begins when a MASCOT activity is entered by the dispatcher and ends when control is returned to the dispatcher is called a SLICE. For the system to work efficiently the lengths of slices taken by different activities must be reasonably equal. Now, in the simplest possible hardware configuration for which

MASCOT is appropriate (a single processor, no external or internal interrupts to activities) the end of a slice can only occur voluntarily. This is the principle of co-operative scheduling. A MASCOT activity has two commands, known as <u>SCHEDULING PRIMITIVES</u>, with which it can terminate a slice. These are:

(a) SUSPEND. The effect of this primitive is to hand control back to the Kernel and cause the dispatcher to schedule another activity which is waiting to run. The suspended activity goes to the 'back of the queue' to await its turn for a new slice.

(b) DELAY (T). This is similar to SUSPEND except that it is guaranteed that the suspended activity will not start a new slice until at least 'T' time units (as defined for the implementation) have passed.

All MASCOT activities are not necessarily equal in respect of queuing for the next available slice. Provision is made for associating a priority with each activity and this may be used by the dispatching algorithm in determining the next activity to run. The Official Definition does not lay down any hard and fast rules as to how this should be done but it does contain some guidelines.

(a)   The scheduling algorithm used must be stated for each implementation.

(b)   Priorities must never be used as the means of achieving correct results. The intention is to provide a way of 'tuning' the system in order to obtain correct response times.

It is appropriate here to mention pre-emptive scheduling as an alternative to co-operative scheduling. This is an acceptable mode of working under MASCOT in which the end of a slice is enforced by a clock interrupt switching control to the Kernel. Responsibility for balancing slice lengths no longer rests with the applications activities and this can lead to added complictions especially with regard to process synchronisation (see next section).

A typical MASCOT system contains a very large number of individual activities. For this reason the standard module, the activity, is not suitable to be the module of control. It would be far too laborious to have to issue commands to initiate or terminate the execution of each activity in the system. It has proved necessary, therefore, to introduce a higher level module for this purpose. The unit of control in a MASCOT system is called a <u>SUBSYSTEM.</u> Subsystems are identified as such during the design process and are implemented during the stage of software construction. They consist of a set of one or more related activities which communicate with each other and collaborate to perform some identifiable task. Dotted lines are used to enclose a subsystem in an ACP diagram.

Figure 6.1 below illustrates the ideas that have been discussed in this section.

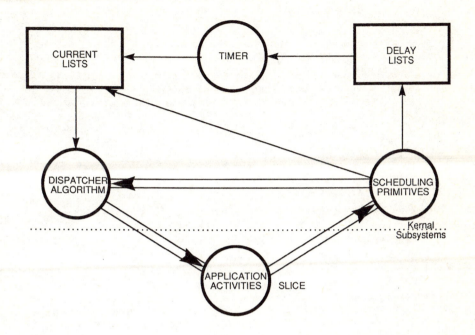

FIGURE 6.1

Here the double lines represent transfer of control and the single lines
data flow. The current lists are queues of activities awaiting their
next slice of action. There may be several such queues in order to cater
for different levels of priority. The delay lists contain activities
which have issued the DELAY primitive and are waiting to be transferred
by the timer to a current list when the specified period has elapsed.
Here again, more than one queue may be needed to handle 'long' and
'short' delays with acceptable efficiency.

7. PROCESS SYNCHRONISATION

The two major problems of process synchronisation which MASCOT sets out
to solve are:

(a) Mutual exclusion of competing processes from a shared resource.

(b) Cross-stimulation of co-operating processes.

Consider first the sharing of a resource between two parallel processes.
The resource in question might be a shared sequence of code which must
only be executed by one process at a time. Naively, one could consider
using a common integer flag to implement mutual exclusion. Of course,
if there was simply a single processor with purely co-operative scheduling and no interrupts, the flag device would be unnecessary. In this
case, each process could be guaranteed uninterrupted access to the
resource. Signalling becomes necessary when another processor is intro-

duced or when any form of pre-emptive scheduling is initiated arising from either internal or external interrupts. In either of these cases, it is possible for the two processes to be simultaneously executing the critical region of code. Moreover, it can be shown that the flag device will not always work and other methods have to be employed, e.g. semaphores (Dijkstra (1968)).

The MASCOT approach to the synchronisation problem is somewhat different: the advantages claimed for it are its directness, understandability and relative simplicity. In place of the semaphore MASCOT uses a CONTROL QUEUE and a writer of MASCOT applications is concerned with the effect of a set of primitive operations which may be applied to a control queue. There are four MASCOT synchronising primitives as follows:

        JOIN   (controlq)
        LEAVE (controlq)
        WAIT   (controlq)
        STIM   (controlq)

JOIN and LEAVE are used in a manner similar to Dijkstra's semaphores to effect mutual exclusion when this is required. Thus:

        JOIN (inq)
                Critical section of code requiring exclusive
                control during access to an IDA
        LEAVE (inq)

where 'inq' is a control queue associated with the IDA interface at which mutual exclusion of activities is necessary. Any number of activities may apply the JOIN primitive to the same control queue, possibly in rapid succession, but only the first is allowed to proceed to the executive of its critical section. The remainder are suspended and held on the control queue 'pending list'. They remain there until the first activity has executed its LEAVE operation. This frees the interface once more and allows the activity at the head of the control queue to enter its critical section.

WAIT and STIM provide the mechanism for cross-stimulation. It is assumed that channels always contain information which allows the state of their data buffers (full, part-full, empty) to be tested. Before inserting or removing data from the buffer an activity checks this state and, if it finds that it is unable to proceed, applies the WAIT primitive to the appropriate interface control queue which it must have previously JOINed. This has the effect of suspending the activity.

The suspended activity will not be in a position to be restarted until the state of the buffer changes as a result of action by some other activity operating at the channel's other interface. It is therefore the responsibility of this second activity to 'alert' the one which is waiting. It can do this by applying a STIM primitive to the WAITing activity's control queue. Thus, a STIM is a software stimulus which can transform any activity WAITing on the control queue to which it is applied from a suspended to an active state.

8.   OPERATIONS IN MASCOT

The preceding sections have attempted to show how the MASCOT methodology enables the user to adopt a modular approach to the design, construction and testing of real-time software. The emphasis has been mainly on the graphical tools, i.e. the ACP Diagram, and it is appropriate here to consider how the graphical design is actually converted into a form suitable for processing by the computer hardware.

The 'Operations' part of the MASCOT acronym is provided by the Kernel, which is a basic operating system. The Kernel LOADS activities into memory, links or FORMS the activities into subsystems, and controls the scheduling of individual activities. The modularity of MASCOT means that activities are specified, designed, coded, compiled and tested as separate modules. The actual code part of an activity is called a ROOT PROCEDURE. An activity is formed when a Root Procedure is connected with actual IDAs. It is possible to set up a library of root procedures, and by connecting certain root procedures in a given way, complex sub-systems can be formed. This is best illustrated by an example which has a military flavour to acknowledge the origins of MASCOT!

9. A MASCOT EXAMPLE - NAVAL AIO SYSTEM

It is required to set up a simple AIO (Action Information Organisation) system which will accept contacts from a Radar subsystem and from a Sonar subsystem. Each contact will be entered into either an air track table or an underwater track table. Finally, details of these tracks can be shown at a VDU and it is assumed that two VDUs are dedicated to displaying UW tracks and one to air tracks. The ACP Diagram of the MASCOT subsystem would appear as shown in Figure 9.1.

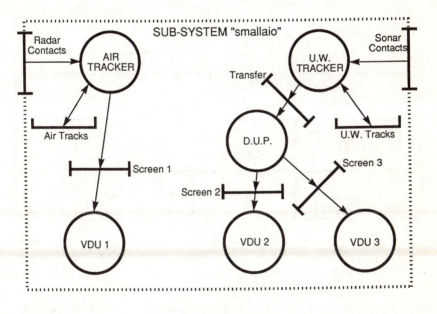

FIGURE 9.1

For the purposes of this example, it is assumed that the air tracker and UW tracker perform identical functions, they differ only in the source of data. The root procedure for air tracker and UW tracker is identical, so only one copy of the root procedure need be kept on backing store.

Similarly, the three VDU activities are identical, so the same root procedure is applicable to them all. A library need only contain three root procedure types, a Tracker, a VDU and a Duplicate (see Figure 9.2 below). These library root procedures are called System Element. Templates or SETs. It is possible to create several copies, or System Elements, in memory. For the example here, two system elements of the Tracker SET, three of the VDU SET, and one of the Duplicate SET are required to make up the ACP Diagram of the subsystem.

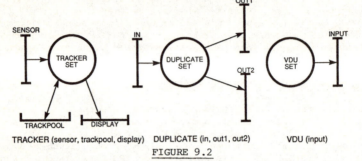

TRACKER (sensor, trackpool, display)   DUPLICATE (in, out1, out2)   VDU (input)

FIGURE 9.2

It is also necessary to set aside areas in memory for the IDAs. The library contains templates of the IDA types. For this example three types of IDA are required: a trackpool SET, a contact channel SET, and a picture channel SET. The MASCOT kernel can be instructed to set aside actual data areas which conform to one or other of the IDA SETs and it can be seen from the table below that eight IDAs are required in total.

IDA sets.	SYSTEM ELEMENTS			
CONTACT CHANNEL TYPE	RADAR CONTACTS	SONAR CONTACTS		
TRACKPOOL TYPE	AIR TRACKS	UW TRACKS		
PICTURE CHANNEL TYPE	SCREEN 1	SCREEN 2	SCREEN 3	TRANSFER

TABLE 9.1

MASCOT sets up or CREATES system elements from system element templates using a CREATE command, e.g.

    airtracker := CREATE (tracker)

Once all the system elements have been Created, and reside in memory, they must be connected to form a subsystem. 'Forming' connects each root procedure element with the actual IDA it will use. To FORM a subsystem called 'smallaio', the Kernel receives a command:

```
FORM smallaio (airtracker (radar contacts, airtracks, screenl),
              uw tracker (sonar contacts, uw tracks, transfer),
              dup (transfer, screen l, screen 2),
              vdul (screenl ),
              vdu2 (screen2 ),
              vdu3 (screen3 ));
```

The flexibility of the MASCOT FORM command can be seen if it is desired
to change the function of 'smallaio' to have two VDUs devoted to the
air picture and only one to the UW picture. A monolothic program may
require extensive rewriting; in MASCOT the same SETs are used to CREATE
the same set of system elements, but now they are FORMed differently,
viz:

```
FORM smallaio' (airtracker (radar contacts, airtracks,
                            transfer),
               uw tracker (sonar contacts, uw tracks, screen3),
               dup (transfer, screenl, screen2),

               vdul (screenl ),
               vdu2 (screen2 ),
               vdu3 (screen3 )) ;
```

The corresponding ACP Diagram is shown in Figure 9.3.

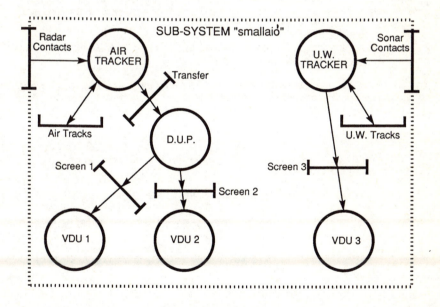

FIGURE 9.3

This section, together with the earlier ones, has given an overview on
how MASCOT attempts to deal with the problems peculiar to real-time
software design. For a more detailed treatment of these issues, the
reader is referred to The Official Handbook of MASCOT (1980).

10. RECENT DEVELOPMENTS - MASCOT 3

All of the material discussed in the preceding sections of this chapter
has referred to the MASCOT definition contained in the 1980 reference.
This particular version is commonly known as MASCOT 2 and is a
descendent of the original version, MASCOT 1, which was issued as an
'Official Definition' document in 1978. The experience of using
MASCOT 2 in practice revealed several weaknesses in its formulation.
This fact, together with the prospect of implementing systems for large,
multi-processor networks and the resultant need for more powerful and
flexible means of design expression, has provided the motivation for the
development of an improved version, MASCOT 3 (MASCOT Handbook (1987)).
It is beyond the scope of the material in this chapter to elaborate at
length about this most recent MASCOT definition. It is sufficient to
say that perhaps the most significant refinement which MASCOT 3 brings
to the concepts discussed here concerns the manner of expressing
network connections. Ideas of an 'access interface' are introduced
using the terminology of 'ports' and 'windows'. The interested reader
is referred to The Official Handbook of MASCOT (1987) for further
information.

11. REFERENCES

1. BUDGEN, D. (1985) "Combining MASCOT with Modula-2 to aid the
engineering of real-time systems" Software: Practice and Experience,
Vol 15, No. 8, pp676-93.

2. DIJKSTRA, E.W. (1968) "Cooperating sequential processes",
Programming Languages, Academic Press.

3. JACKSON, K. and SIMPSON, H.R. (1975) "MASCOT - A modular approach
to software construction, operation and test", RRE Tech. Note No.778.

4. PHILLIPS, C.S.E. (1967) "Networks for Real-Time Programming",
Computer Journal, Vol 10, No. 1, pp46-52.

5. The Official Handbook of MASCOT (1980), MASCOT suppliers
Association, RSRE, December.

6. The Official Handbook of MASCOT (1987), Computing Division,
RSRE Malvern.

Chapter 10

Software fault tolerance

S. Bennett

1. INTRODUCTION

Methodologies for analysing the reliability of complex systems and techniques for making such system tolerant of faults, thus increasing the reliability, are well established. In hardware the emphasis is on improved components and construction techniques and the use of redundancy to maintain critical systems functions in the event of single failures (in some systems multiple failures). In software the emphasis has been on improved software construction techniques - "Software Engineering" - to reduce latent errors; there has also been work on techniques to introduce redundancy into software systems.

The majority of work on fault tolerance has concentrated on what Anderson and Lee have termed "anticipated faults", that is faults which the design can anticipate and hence "design" in tolerance. A much more difficult an insidious problem is that of faults in the design of the system. These are by definition "unanticipated faults" (and unanticipatable). Design faults can occur both in complex hardware and software but are more common in software and much of the effort of software engineering has been directed towards reducing design faults, that is unanticipated faults.

The basic ideas of reliability have been expounded in several text books (see for example Smith 1972) and a very brief review is given in the next section.

2. RELIABILITY DEFINITIONS

The theory of reliability is based on probabilistic concepts and reliability R is defined as the probability that a system will perform a required function, under stated conditions, for a stated period of time. For systems with a constant failure rate z the reliability R decreases exponentially with time according to the relationship

$$R = e^{-zt} \qquad (0 < R < 1)$$

Mean Time Between Failure (MTBF) is a commonly used measure defined as

$$MTBF = \int_0^\infty R(t)dt$$

which for a non-redundant system reduces to the reciprocal of the failure rate thus

MTBF = h = 1/z

In practice electronic equipment follows the so called bath tub curve and the failure rate is constant only during what is termed the "useful life" phase. At the end of this phase the failure rate grows rapidly and hence their is a rapid decrease in reliability.

Unavailablity is used in systems which are repairable following all failures. Unavailibity U is a measure of the time which the system is out of service and is given by

$$U = MTTR / (MTBF + MTTR) \quad (0 < U < 1)$$

where MTTR is the Mean Time To Repair. Unavailability is sometimes expressed as Forced Outage Rate (FOR) and is expressed as a percentage. As an alternative to unavailability, "availability" A is used $(U + A = 1)$.

A common technique to increase the reliability of a system is to have reserve or stand-by equipment which is brought into use when the main equipment fails. For a system or component with a failure rate for unrevealed faults z' and a mean time to test of T (assumed small) then failure on demand (FOD) is given by

FOD = z'.T/2

3. USE OF REDUNDANCY

A well established technique for increasing the reliability of hardware systems is to duplicate or triplicate hardware units. Hardware redundancy can be introduced in two forms characterized static (or masking) redundancy and dynamic redundancy. In static redundancy the duplicate components are such that they take over the operation in the event of failure and hence mask the failure of the system from the environment. To detect that a failure has occurred some secondary indication of failure is necessary. In dynamic redundancy an unit is used to detect an error which must be corrected by redundancy elsewhere in the system. For example a computer control system with several processors operating in parallel and carrying out (or capable of carrying out) the complete functions of the system with the final action being determined by a majority voting system involves static redundancy. An example of dynamic redundancy is a distributed computer system in which each unit carries out a specific subset of the overall functions but which has some means of detecting when a unit fails and transferring the actions of the failed unit to one or more of the operating units.

The most common example of the static redundancy approach is Triple Modular Redundancy (TMR) in which three identical units are run in parallel and a majority voting system is used to check the outputs. The systems is designed to tolerate failure of a single module as it produces output only when two modules agree (2-out-of-3) system.

Although the concepts of multiple channels and majority voting are simple there are some serious practical problems. An obvious problem is that a parallel channel system does not provide any protection against a common mode failure. For the TMR system shown in figure 1 if identical copies of the software are used in each channel then a software error will cause a common mode failure.

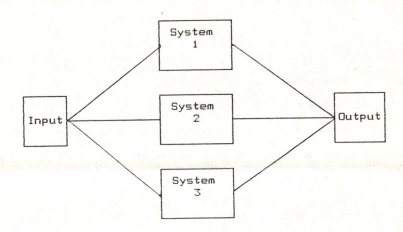

Figure 1 Triple modular redundancy

If the signals entering the majority voting unit are two valued logic signals then a clear unambiguous result is obtained. If any two are true the output will be true, if any two are false then the output will be false. However, if the signals are modulating signals then the majority voting unit is more difficult to implement. Three common techniques are:

Average or Mean value: this is very simple to implement and can be satisfactory if the signals do not deviate from each other greatly or if a large number of channels are used. For a TMR system a fault which causes one of the signals to go to a maximum or minimum value can introduce a significant error in the output.

Average with automatic self-purging: the average is calculated but if one of the channels deviates by more than a specified amount from the average that channel is disconnected. This technique can generate a large transient when the faulty channel is disconnected.

Median Selection: instead of taking the average the median value of three or more channels is selected. This technique avoids large transients.

In the above it should be noted that only the averaging with automatic purge provides detection of failure of a channel.

In any system employing parallel channels there will be divergence, that is the individuals channels will not give identical values and further more the differences in values may change with time without implying failure of any particular channel. To avoid such divergence resulting in disconnection of a functioning channel some form of channel equalization is necessary (see Ham 1984).

4. FAULT TOLERANCE

As indicated above one of the basic techniques for introducing fault tolerance is to add redundancy to the system. However, care is required as adding redundancy adds complexity to a system and complexity can increase the likelihood of faults occurring. Thus great care has to be taken in adding redundancy to a system.

Dealing with reliability and fault tolerance of mixed hardware/software systems introduces difficulties in determining the actual reliability structure of the of the system. Put in simple terms the software is distributed over several hardware elements, for example, a software module driving an input device is distributed over the CPU, memory, interface bus and the input device. Thus although at the software design stage the input driver is (or should be) clearly defined as a module this structure may not be apparent when the software is run.

Anderson and Lee propose a model in which the hardware is viewed as maintaining an interface on which the software is executed. They term this interface the interpretive interface and they represent the system C by model

$$i \quad \frac{S}{H}$$

where if C is a computing system then S is the memory containing the program (code and data) and H is the rest of the hardware. The interpretative interface i represents a language that provides objects and operations to manipulate those objects. At the simplest level this is the machine language of the hardware.

The model is extended in two ways: one is a multilevel interpreter system which can be represented as

$$\text{BASIC code (keyed)} \quad \frac{\text{BASIC program}}{\text{BASIC interpreter (and operating system)}}$$
$$\text{machine code} \quad \frac{}{\text{hardware}}$$

The other is the concept of an extended interpreter represented as

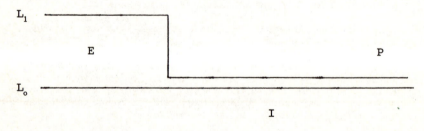

This model assumes that some interpreter (language) L_o is available but a system is required to support an interpreter L_1. If L_o contains some of the facilities required by

L_1 then it is not necessary to provide a complete interpreter, an interpreter which uses some of L_o but extends it to provide extra facilities can be used. The program can then be considered as being in two parts, part E which is written in L_o and part P which is written in L_1. The interpreter extension model clearly can be used to represent operating systems. It can be extended to represent multilevel extended interpreters.

In using this modelling approach to further concepts are helpful. They are used to distinguish between facilities provided by the interpreter and those which are programmed in the interpreted system:

*   a mechanism provides a specific facility at the interpreter interface and is implemented as part of the interpreter;

*   a measure performs some specific task and is implemented by means of a set of instructions in the system that is interpreted.

Consider the input of data into a program written in BASIC. Suppose the program has to accept integer values. It can be written using the standard input statement with a BASIC integer variable given as part of the input statement. In this case if the user inputs alphabetic characters the BASIC interpreter contains a mechanism that will detect an error and will force an entry to the error handling system. The system designer will thus have relied on an interpreter mechanism. An alternative would be to write a measure to check input; by reading the input into a string variable and perform the string to numeric conversion by program the input checking would be performed using a measure not a mechanism.

Mechanisms are likely to be more general purpose and widely applicable than measures and Anderson and Lee place emphasis on generating fault tolerance through the use of mechanisms.

A second basic idea is to separate normal behaviour of a system from abnormal behaviour. It is assumed that a fault condition will generate an error which will be an exception. A way of reducing complexity is to separate the response to an event into a normal response and an exception response. System designers and implementers should be able to deal with the two responses separately.

For example a programmer does not normally have to deal with check for "divide by zero", the computer hardware is designed to detect this fault and generate an exception response. The normal exception response in this case is a software trap which transfers control to an error handling routine. In many systems the error handler has a fixed predetermined response - in this case typically to halt execution of the program. For real-time systems the fixed response is often not wanted and if it cannot be disabled the designer is forced to insert additional coding to trap errors prior to them initiating the built in mechanism.

What is required is a system which allows the designer to explicitly invoke an exception mechanism in response to some predetermined event. This is generally referred to as rasing an exception. Facilities for enabling and disabling particular handlers are also needed.

4.1 Fault Detection Measures

4.1.1 Replication checks

The use of replication checks involves duplicating or replicating the activity of the system that is to be checked. A typical example is the use of dual processors each of which runs the system software. The actions of the two systems are compared at specific times or in response to specific events. If it is assumed that the system design is correct and that failures will be independent then no underline fault will go undetected. With replication some additional operation is necessary to determine which of the two units is faulty.

The procedure can be extended to include triple redundancy (TMR) and multiple redundancy (N-Modular Redundancy, NMR). Higher orders of redundancy do not increase the error detection capability but they enable the faulty unit to be identified and the system to continue to run.

Replication is expensive and can increase the complexity of the system. The expense and complexity can be reduced by limiting replication to selected critical functions. The limitation of replication using identical units is that it cannot detect common mode faults due say to errors in the design of the system. In order to check for design faults the duplicate system would have to be a completely independent design. Adopting such a technique is costly. An alternative (Ayache et. al.) is to use a model of the system: they suggest using a Petri-net based model to check the high-level control flow of the operational system.

Replication can be used in a simple way to detect transient errors by repeated operation of the same system. This technique is frequently used in communication systems and in software for accessing disc drives were several attempts will be made to say read or write to a disc before reporting an error and halting the action.

4.1.2 Expected Value Methods

The majority of methods used to detect software errors make use of some form of expected value. The correct progression of a program through a sequence of modules can be checked by using 'baton' passing. Each module passes to its successor module a unique numerical value. If the value received by the successor module does not match the expected value then the sequence of modules is incorrect.

Another technique, sometimes referred to as an 'assertion test' is to build into the program logical tests. Range checking and input-output checks offered in many modern compilers automatically insert coding for assertion testing. Assertion tests can also be incorporated into the application program explicitly by the designer. For example, if on the basis of the software specification, an array or table should at some particular point contain only positive values. Then a test to check that this is the case can by inserted in the program code.

Because of the overhead which this type of checking incurs it is often restricted to the testing period. Many compilers provide switches to allow the range checking and other forms of checking to be selected at compile time. Similarly other forms of checking code can be put into conditional compilation segments to allow it to be omitted in the delivered software system. Careful use of assertion testing on entry to and exit from modules can help to stop a fault generated in one module giving rise to faults in other modules

Adoption of a 'good' module division at the design stage enables 'defensive programming' to be used check data passing between modules.   It is strongly recommended that data passing between modules be carefully checked since if the design has been well done the properties of data at module boundaries should have been carefully described and documented and hence checks should be capable of giving a clear indication of any faults.

4.1.3 Watchdog Timers

Timing checks, that is checks that some function has been carried out within a specified time can be used to detect errors in a system.   It should be noted that timing checks do not indicate that a system is function correctly they can only indicate that it has failed in some way.

Watchdog timers are commonly used to monitor the detailed behaviour of a real-time system.   The watchdog task 'watchdog_timer' is run at fixed time intervals as a high priority task.   It decrements counters (timers) held in the pool and checks that the counters have not reached zero.   If a counter has reached zero the watchdog_timer task signals to the executive that a time-out has occurred.   The tasks T1, Ti, Tn being monitored by the watchdog periodically reset the counters thus a time-out occurs when they fail to reset the counter within a given time interval.

Watchdog timers can be used to monitor peripherals.   By monitoring critical tasks they can also show up a temporary overload on system, or clearly show that a system has insufficient processing capacity.

The watchdog timer can also be applied to software communication channels.   Problems of synchronization can often be simplified if a task can restrict its commitment to waiting for information for a predetermined time.   In many real-time applications there are circumstances where it is better to continue with out-of-date information (for example the previous value) or to estimate a data value rather than simply to wait indefinitely.   For example in a feedback control system it is normally better to continue, in the event of a failure of a single instrument, with an estimated value than to halt the controller action.   The failure must of course be reported and the instrument repaired or replaced within a reasonable time.

4.1.4 Reversal checks

For systems in which there is a one-to-one relationship between input and output a useful error check is to compute what the input should be for the actual output and compare the computed input value with the actual input value.   For example disc drivers often read back the data segment which has just be written and compare the data read back with that which was sent.   Mathematical computations often lend themselves to reversal checks, and obvious example is to check the computation of a square root by squaring the answer.   However, care is needed in such checks to take into account the finite accuracy of the computation.

A variation on this technique can be used in systems in which there is a fixed relationship between the output and input.   For example for matrix inversion the product of the input matrix and the output matrix should be the unit matrix.

4.1.5 Parity and Error Coding checks

Parity and error coding checks are well known techniques for detecting (parity) and detecting and correcting (error codes) specific types of memory and data transmission errors. The most widely known is the single bit parity check applied to memory storage and asynchronous data transmission which will detect the loss of an odd number of bits in a storage or transmission unit. A simple parity is an effective and efficient solution to detecting errors when they occur as single bit errors in a unit; it is not an effective method if typical failures result in multiple-bit errors.

More complicated codes such as the Hamming, cyclic redundancy, and M-out-of-N codes are used to detect multiple-bit errors and some of these codes allow reconstruction i.e. correction of the error. The use of these codes involves adding a greater amount of redundancy than the use of a simple parity code.

The above codes are generally used to detect errors due to hardware component failure. Codes can also be applied to data to detect software errors. For example the use of a checksum computed on a block of data and held with it can be used to detect both hardware and software errors. A combination of parity and checksum added to a data unit and a block of data respectively provides an efficient method of error detection for large and complex sets of data.

4.1.6 Structural checks

Checksum codes are typically applied to data structures and provide a check on the consistency and integrity of the data contained in the structure. Errors can also occur in the data structure itself, for example pointers in linked lists can be corrupted. Thus checks on the structural integrity are also required.

A very simple check in a linked list is to maintain in the list header a pointer to the last item in the list as well as one to the first item. If the pointer to the last item does not correspond to the last item identified by following through the chain of pointers from the first item then an error has occurred. A simple check of this form, however, provides little information on the error and the possible damage. Also there is little information to help with recovery. An frequently used alternative scheme is the doubly linked list in which each element contains both a forward and backward pointer.

4.1.7 Diagnostic checks

Diagnostic checks are used to test the behaviour of components used to construct the systems, not the system itself. The checks use a set of inputs for which the correct outputs are known. They are usually programs used to test for faults in the hardware of the system and typically are expensive in terms of the time and resources used. As a consequence they are rarely used as primary error detection measures but are restricted to attempts to locate more precisely faults detected by other means. A difficulty widely experienced is that it is not easy to design diagnostic checks that impose on a component conditions as stringent as those imposed by the actual system.

4.2 Fault Detection Mechanisms

Some mechanisms for fault detection based on interface exceptions have already been mentioned. These include illegal instruction, arithmetic overflow and underflow, protection violation and non-existent memory. Few systems offer mechanisms beyond these. Although, for example, a compiler for a strongly typed language will detect and flag as errors attempts to perform arithmetic operations on variables declared of type character, the underlying hardware will not distinguish between words holding integers, characters, reals, etc. Hence a run time fault that results in an attempt to add a word containing a character to a word containing a real would not result in an exception being signalled. One mechanism to overcome this problem, the use of tagged storage has been offered on a few computer systems. With tagged storage a few extra bits in each word are allocated to be used to identify the type of object being held in the word.

Other useful mechanisms for fault detection are ones which detect attempts to exceed array bounds or to exceed ranges specified for integers and these are provided in a few computers.

4.3 Damage containment and assessment

Damage results as the effects of faults or errors propagate throughout the system. A major problem in assessing the extent of the damage arises through the presence of parallel operations. This is true for both standard sequential programs running on a single processor and for multi-tasking and multi-processor distributed systems. The larger the number of tasks and processors the greater the problem.

One approach to the problem is to use techniques similar to those used for preventing clashes between competing concurrent processes. Sections of code are treated as critical sections, or certain objects are only permitted to belong to one tasks at any particular time. Access control mechanisms - monitors, guards, locks - are used to protect critical sections or objects.

Obvious containment measures are to limit access to a file while a task is updating. If checks on data consistency and structural checks are carried out prior to releasing the file then effects of a faulty update can be prevented from spreading.

An important technique for assessing damage is to mark bad data in some way. This technique can also be used for containment if incoming data to a task is checked to see if it is marked as bad. In distributed systems it is important to test thoroughly all incoming data messages since the detection of errors or bad data at the input can prevent damage spreading between processors.

4.4 Fault Recovery Measures

4.4.1 Check Points

A standard technique which has been used for a long time in data processing systems is to insert check points in the program. At a check point a copy of data which would be required to restart the program at that point is written to backing store or to some protected area of memory. If a fault is detected at a test point or the next check point in the program then the program is 'rolled back' to the previous check point and restarted.

The strategy can be readily adapted for use in real-time systems. In a control system for example back-up copies of values such as set-points, controller parameters and possibly a 'history' of selected plant variables are held either on backing store or in battery backed-up memory. In the event of failure they can be reloaded.

Recovery mechanisms involving the storage of data values can provide protection against hardware faults or software faults which do not re-occur. However, in the majority of cases if a software module fails once it will contain to fail. The only way in which the system can continue is if a replacement module can be activated. This technique implies redundancy.

4.4.2 Redundancy

A standard hardware technique for increasing reliability and providing for error recovery is the introduction of redundant modules. Similar techniques can be applied to software. There is, however, a fundamental difference between the use of redundancy in hardware and in software. In hardware it is assumed that the each unit functions correctly according to its specification and that failure is due to some physical cause - wear, a faulty component, unusual physical stress - that is not common to both units.

In software the major causes of failure are design faults - ranging from misinterpretation of the specification to simple coding errors. Simply replacing a software module which has failed with an exact copy will result in an immediate failure of the replacement copy. The approach required for software redundancy is more complicated requiring the production of independent software modules from the same specification.

The production of independent software modules is not a simple task. It is not sufficient to use a completely separate and independent team of designers since some coordination is required to ensure that they do not choose for example the same algorithm for a crucial calculation.

A method for introducing redundant modules into a system is the use of <u>recovery blocks</u>. Associated with a checkpoint is a recovery block which contains one or more alternative code modules. If an error is detected the primary alternative is used. At the end of the recovery block there is an acceptance test, if, as a result of running the alternative code module the acceptance test is passed the recovery block is exited. If the acceptance test fails then the next alternative module is tried. If all code modules are exhausted before and acceptable result is achieved then an error is reported.

The general structure is

> *establish recovery point*
> *primary module*
> *acceptance test*
> *alternate module*
> *acceptance test*

and this is normally expressed using the syntax

> *ensure*   *<acceptance test>*
> *by*   *<primary module>*
> *else by*   *<alternate module 1>*

```
else by        <alternate module 2>
  .
  .
  .
else by        <alternate module n>
else error
```

4.5 Special Features of Real-Time Systems

4.5.1 Deadline Mechanisms

The provision of fault tolerance for real-time computer systems has to take into the account that the time to perform some operation enters into the specification of the system and failure to complete in some specified time constitutes an error. Therefore there may not be time to carry out some of the recovery procedures outlined above unless special techniques are used. One such technique is the deadline mechanism.

This is an attempt to deal with the requirement that service must be completed within a specified time. The problem is illustrated in the following code segment for a fault tolerant navigation system.

```
every         second
within        10 milliseconds
calculate by  read sensors              "primary module"
              calculate new position
else by       approximate new position from
              old position               "alternate module"
```

The *every* statement is used to specify the repeat time of the particular task. The *within* statement specifies the maximum amount of time that can be permitted to elapse between starting the task and getting the results back from it. Two modules are specified, a primary module which reads the sensors and calculates the new position and a single alternate module which estimates the position from the previous position value. It is assumed that the alternate module is error free and requires less time to run than the primary module.

In order to meet the overall time deadline the system implementor has to accurately determine the execution time for the alternate module in order to determine how long cam be allowed for the primary module to produce a good result. For the 10mS deadline if the alternate module is estimated to take 3mS then the primary module must return a result within 7mS.

A characteristic of real-time systems that can be used to simplify the problem of fault recovery; this is that tasks are repeated at frequent intervals with new data sets supplied at the system boundary. For example for a feedback control loop a single set of bad data values can be ignored if it can be assumed that the next set of readings will be correct. If the control loop is stable and well designed the effect will be minor. The effect of a single bad set can be reduced further by replacing the bad set with a predicted value based on some form of extrapolation. The use of predicted values can allow for a series of bad values until some other corrective action can be taken, for example switching to another instrument.

A simple recovery block could be

```
ensure              <data good>
by                  <normal value module>
else by             <predicted value module>
else error
```

A more complex error recovery block would need to take into account in the acceptance test the time for which the data was bad, that is how long is it acceptable to use predicted data. And also in practice care would be required to avoid generating a spurious error if the next good data value had diverged from the predicted value.

4.5.2 Bad Data Marking

The use of modular design techniques facilitates the use of defensive programming to detect bad or faulty data. The major problem arises not in detecting bad data but in deciding what action to take. An interesting solution to this problem is that adopted by the designers of the CUTLASS system. Associated with each data variable is a tag bit which indicates whether or not the data is bad. When a test on the data is carried out the tag bit is set or reset according to the result of the test. Bad data is allowed to propagate through the system but carries with it an indication that it is bad. The language support system contains known rules for evaluating expressions containing bad data. For example if two logic signals are combined using the OR function then both have to be bad for the result to be tagged as bad, whereas for the AND combination if either signal were bad then the result would be tagged as bad. With this approach the program module which determines that a data value is bad does not need to know which modules it should inform since the information is automatically conveyed with the data.

CUTLASS was designed to support distributed control and the use of tagged data avoids some of the problems of synchronizing recovery in distributed systems. The tagging of data allows easy implementation of recovery blocks and gradual degradation of the system as a particular task can be programmed to select alternate function modules according to the status of the input data set.

For example if the control signal calculated for a particular actuator is tagged as bad the output module can be programmed to leave the actuator set at its last value, move the actuator to a predicted value, or move the actuator to a predicted value unless the bad data time exceeds a preset amount at which time the actuator position is frozen. In the recovery block notation this can be expressed as:

```
ensure              <data good, timeout=false>
    by
    ensure          <data good>
    by              <primary module - output data>
    else by         <predict output>
    else error
else by             <do not move actuator>
else error
```

The module in the inner block used to predict the output would need to have an acceptance check in it to avoid large movements of the value if the bad data indication came at a time when a large value movement had been occurring. Such an acceptance check could force an error which would cause the outer block to institute the do not move the actuator module.

5. CONCLUSIONS

We have described some of the problems of designing and producing reliable software and how these problems are different from those associated with hardware reliability. The importance of being able to produce reliable and fault tolerant software will continue to increase as the use of devices incorporating computer processors grows. Improved production methods will increase software reliability, however, it will remain difficult to make software based systems tolerant of unanticipated errors. Safe failure of systems incorporating software will continue to depend in the last resort on hardware safety provisions.

6. REFERENCES AND BIBLIOGRAPHY

Anderson, T. and Lee, P.A., *Fault tolerance, principles and practice*, Prentice Hall, Englewood Cliffs NJ, (1981). This book contains an extensive bibliography.

Anderson, T.(editor), *Resilient computing systems*, Wiley, New York, (1985).

Anderson, T., Randell, B. (editors), *Computer systems reliability*, Cambridge University Press, Cambridge, (1979).

Ham, P.A.L., "Reliability in computer control of turbine generator plant" in Bennett, S., Linkens, D.A.,(editors) *Real-time Computer Control*, Peter Peregrinus, Stevenage, (1984).

Myers, G.L., *Software Reliability*, Wiley, New York, (1976).

Nelson, V.P., Carroll, W.D., *Tutorial: fault-tolerant computing*, IEEE Computer Society Press, Los Angeles, CA, (1987).

Shrivastava, S.K. (editor), *Reliable computer systems: collected papers of the Newcastle Reliability Project*, Springer-Verlag, Berlin, (1985).

Chapter 11

An OOD methodology for shop floor control systems

N. K. Stanley

1 INTRODUCTION

This article will describe how object-oriented design
(OOD) was used to produce a high-level design in
conjunction with a top down stepwise refinement
methodology. It will also describe how object-oriented
design was used recursively through several layers of
recursion on the design of a large complex software
system.

2 OBJECT-ORIENTED DESIGN

2.1 Overview

The term 'object-oriented' originated in work on the
Smalltalk programming system. This system is based on the
idea of objects representing a given process with a
communications ability to the external world. The object
is however protected from external viewing or influence.
Data within an object can only be accessed by the
operations embedded within the object.

My special thanks to Genie King and
the Montpellier design team.

There are four major themes to object-oriented design

- Object

    | Objects encapsulate data and the operations that can be performed on that data. Formally known as encapsulation.

    | An operation on an object is executed by sending a message requesting the operation to that object.

    | An object is an instance of a class which describes common structure and behaviour of all members of that class.

- Class's

    | Classes are organised hierarchically thus allowing inheritance.

- Abstract data types

- Inheritance

OOD allows behaviour to be determined at run-time, known as late binding.

There are many ways to choose the objects that make up a design. One method is to make the objects represent the components of the application problem. In this high level design we took the problem to be the Product Functional Specification (PFS). Decide what objects or entities the user will be dealing with in the application and define what operations are allowed against each object. Objects typically contain other objects and this soon became extremely evident in our development. This part of this design is crucial, extremely difficult but will repay in time saved in later development stages.

2.2 Object-Oriented Design Approach

Object-Oriented Design is based on a number of different sources, one of the main ones being the article by Parnas, it argues the case where the modules of a system are designed in such a way that separate groups can develop them providing a high level of independence. The interfaces revealing as little as possible about what actually goes on within the module itself.

The Parnas approach has 4 major advantages:

-   Concurrent work - leading to a shorter development cycle

-   More flexible product

-   Due to abstraction, the modules can become more comprehensible

-   Enables correctness and validation to be performed more easily.

The steps are basically:

-   Define the problem

-   Define an informal strategy for solving the problem

-   Define a formal strategy

    | Identify the operations on the object

    | Establish the interfaces to each kind of object

    | Implement the internals of the operations

The first step was to take the Product functional specification (PFS) and functionaly decompose the specification into unique segments. The PFS describes the external characteristics and functions as they will be seen by the users of the product.

The second step is to define the problem in terms of its objects and operations at its highest level in terms of the objects and operations that are quoted in the product functional spec (the problem requiring a solution).

The third step (this involves sub-steps) is to formalise the strategy decided upon. Identify the object operations and the data types within the. object. We used the heuristic technique of:

-   A common noun suggests a type.

-   Noun or direct reference suggests an object.

-   A verb, attribute, predicate, descriptive expression suggests an operation.

A sub-step was to define the interfaces, the part seen by the external world and the method of communicating with the object (the external view). Only the information required by a user of the object was shown. The next sub-step was to create the 'hidden' bodies of the objects. In effect a top-down object-oriented design. This was the approach used to produce the high-level design for this CIM software product.

3 HIGH-LEVEL DESIGN PROCESS

3.1 Overview

The following diagram shows the steps in the high level design process as defined by the CIM development team.

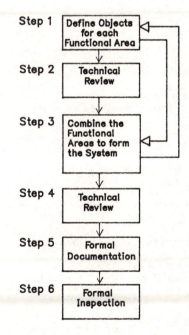

3.2 Step 1: Define the Objects for Each Functional Area

An object is defined as data and the operations that act upon it.

The Product Functional Specification document has defined within it all the functions of the CIM program product. This document will be used to define the starting set of objects. Each functional area will be analysed and the objects in each area will be identified. This will involve interaction with other functional areas to determine how the areas interact and link with each another.

This step will provide the first management checkpoint and will supply a clear set of tasks for division amongst the development team. It will also supply a better understanding of the complexity of the product and the resources necessary to complete the project.

3.2.1 Output for Step 1

The output for Step 1 will consist of:

-   Assumptions, a statement of all assumptions made by the designer when producing the high-level design document.

-   Object Diagram.

    | Text description of what the objects domain is and its functions.

    | Object Diagram, high level view of the objects domain.

-   Diagram of All objects referenced within this HLD

    | Diagram showing all the objects referenced, denote the "common" objects by placing a "*" in the object circle.

    | Text description of each object and the operations that the HLD requires each object to perform.

Split the list of objects into TWO alphabetical
ordered lists. The first list to contain objects
that are "common" or not supplied by your own
functional HLD. The second list to contain the
descriptions of all the objects that your own HLD
will produce.

- Object Flow Diagrams.

  | Text description

  | Object Functional Diagram (OFD)

  | Behaviour text for the OFD

- Data Usage Table

- Panel Usage Table

- Application Program Interface (API) Table

  | A list of published API's is in section 3 of the
  PFS.

  | A summary list of all the API's is required at the
  start of this section.

- Internal Program Interface (IPI) Table

  | A summary list of all the IPI's is required at the
  start of this section.

3.2.2 Object Diagram - Overview

The initial diagram should illustrate the overview of the
functional area (as mapped to the PFS), where all
operations on the object are shown.

The main objects are illustrated by containing them within
a large circle (representing the boundary of the
functional area). The operations acting on this functional
area are illustrated by boxes containing the names of the
operations. The operations are connected to the functional
area by attaching the box to the functional area circle.

If there are any IPI at this level, highlight the IPI
operations, by placing a '@' symbol in their box's.

The accompanying text will give an overview of the functional area and a description of the operations that act upon this object.

3.2.3 Object Diagram Narrative - An Example

The Transport Interface (TI) is used to initiate a move between workcells and within a Workcell. TI is an interface between the Workcell and a Move System Shell. The following operations can be performed on the T.I. functional area:

```
Move Request
Move Complete
Move Cancel
Move Status
```

3.2.4 Object Diagram Overview - An Example

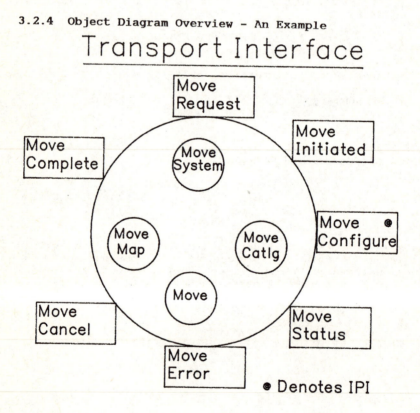

Transport Interface

Move Request

Move Complete

Move Initiated

Move System

Move Configure ●

Move Map

Move Catlg

Move Cancel

Move

Move Status

Move Error

● Denotes IPI

3.2.5 All Objects Referenced

An object diagram will show all the objects that have been
identified for a functional area. An object is visually
represented by the use of a circle, the name of the object
is contained within the circle.

Structure the layout of the diagram such that it shows a
'layering'.   The top 'layer' should show the first level
objects, the next 'layer' the second level etc.

An '*' placed in the circle indicates that it is either a
"common" object or an object provided by another
functional high-level design (HLD) area.

The accompanying text will briefly describe each object.

The descriptions have been split to clearly indicate those
that are produced by other functional HLD areas. This will
help in the combine object phase.

3.2.6 All Objects - An Example

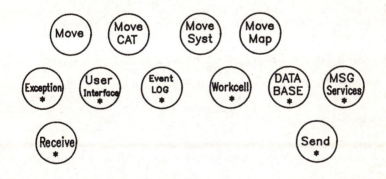

3.2.7   Objects Referenced Description - An Example

MOVE CATALOGUE:

The Move Catalogue object allows the entire set of outstanding moves to be viewed. Each node is responsible for its own Move Catalogue, ensuring a distributed approach and an easier recovery situation.

The following options are required :

    Add Move
    Delete Move
    Update Move
    Read Move

3.2.8   Object Flow Diagrams (OFD)

Following this 'total view' diagram there should be a series of flow diagrams. These will illustrate the objects (and their sequence) referenced by each operation on the functional area. In effect these will be detail explosions of the 'total view'.

Each OFD should only show one level (or layer) of detail, this is to ensure a structured approach to the design.

Two sets of narrative are required for each object flow diagram, a brief description plus a behaviour narrative.

The description text should indicate what the flow diagram is trying to achieve. The object flow behaviour narrative will briefly outline how the objects are used and/or interact with each other.

The behaviour narrative should be worded so that any person can read and understand. The following simple set should be used:

    - IF...ELSE...ENDIF
    - BEGIN...END
    - DO...ENDDO
    - DO WHILE...ENDDO
    - REPEAT UNTIL...ENDREP

```
- CASE...WHEN...DEFAULT...ENDCASE
- EXIT
- /* comments...description */
```

When using an object, use the convention 'call OBJECT(ACTION)' this way the Behaviour code can be matched to the OFD and the API/IPI data.

3.2.9 OFD Standards and Conventions

A circle containing the name of the object is used to illustrate an object.

An arrow is used to connect one object to another.

A box drawn touching the apex of the circle illustrates the operation being performed on the object. The box contains the name of the operation being performed on the object.

An arrow starting from a small solid circle inside an object (circle) and linking to an external object (circle) indicates that the external object(s) is contained within that object. In effect this is illustrating a detailed explosion.

3.2.10 OFD Description - An Example

The T.I. object receives a Move Request (MR). The operation then 'explodes' within the T.I. object, the Move Map object is called to verify the operation message (MR API).

If the MR is valid then the Move is added to the Move object is called.

The Event Log object is now called (if logging has been configured by the user) and this object adds a copy of the Move Request to the event log.

If the linkage is a Move System object then the MR is forwarded to the Move System. Upon acknowledgement of the Move Request, the status of the Move System (within the Lafayette system) is updated.

3.2.11   OFD - An Example

Transport Interface: Move Request Operation

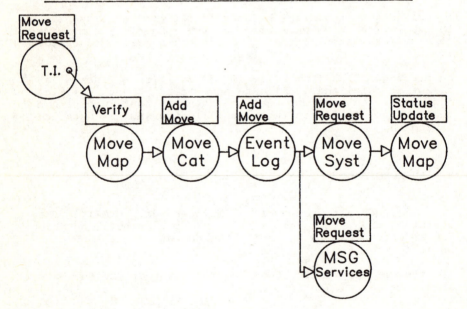

3.2.12 OFD Behaviour - An Example

```
/* Move Request received for processing                     */
/*                                                          */
/* Pass the required information to verify the following:   */
/* a)Move Request reference points exist and are valid      */
/* b)a Move system has been defined between these two points*/
/*   (user shell, manual ...)                               */
/* c)a Move Initiated has been configured to this move      */

call MOVE_MAP(VERIFY)

IF move request verified

    /* Move Request is added to the catlg of outstanding move*/
    /* requests along with the configuration data requested. */

    call MOVE_CATLG(ADD_MOVE)

    IF auto mode

        /* Issue the Move Request to the MOVE SYSTEM SHELL    */

        call MOVE_SYSTEM(MOVE_REQUEST)


    ELSE its a manual link

        /* Issue a message to tell the supplier Workcell that */
        /* he has a Move Request waiting.                     */

        call MSG_SERVICES(MOVE_REQUEST)


    ENDIF

ELSE move request not configured

    /* Move Request Rejected                                 */
    /* returned by TI.                                       */

ENDIF
```

3.2.13   Data Usage

All data in an object will be identified in this section, which contains the logical name of the data, the object which encapsulates this data, and the purpose of the data. Descriptive information on each data item will also be supplied.

3.2.14   Panel Usage

All panels associated with an object will be identified in this section. The section will contain the name of the panel, the object associated with this data, the purpose of the panel, panel name and data on the fields displayed on the panel.

3.2.15   Identification of Application Program Interfaces (API's)

An API is a well defined and published interface. The API is used for passing information between a) external shells and internal programs/functions, and b) internal programs/functions and internal programs/functions.

This section should be headed by a list of all the API's contained within the section, this will enable a cross-reference to be made between the PFS and the HLD document.

This section identifies all API's which can be used by both shell and Internal programs to access Internal functions and data on the fields contained within the API.

3.3   Identification of Internal Program Interfaces (IPI's)

This section should be headed by a list of all the IPI's contained within this section, this will help at the Combine Objects phase.

An IPI is a well defined internal interface. The IPI is used for passing information between internal programs/functions and internal programs/functions only.

An IPI can not be used to communicate between shells and Internal Programs/Functions. The information required in this section, is identical to that required in the API section.

3.4 Step 2: Technical Review

3.4.1 Objectives of Technical Review

To ensure that the HLD fulfils all the functions contained within the published PFS.

To verify the correctness of the design and internal consistency of the application development design.

To ensure that all the objects interface correctly and provide the required operations and functions.

To ensure that the API's specified are consistent and documented.

To ensure that all functional areas provide a well documented, readable and under-standable High Level Design document, for subsequent usage in Low Level Design (LLD).

3.4.2 Information Required for a Technical Review

The following is required:

- Object's Diagram
- Object's Diagram text description
- Object Functional Diagram
- Object Functional Diagram text description
- Object Flow Diagrams
- Object Flow Diagrams text description
- Object Flow Diagrams behaviour description
- Data Usage Table
- Panel Usage Table
- Application Program Interface (API) Table
- Internal Program Interface (IPI) Table

3.4.3 Technical Review Attendees

Mandatory requirements for a review team are:

-   Knowledge of Process

-   Knowledge of OOD

-   Knowledge of Box Structured Methodology (BSM)

Technical Reviewers, must be familiar with the functional area as described in the PFS.

The objective of each Technical reviewer is to ensure that all API and function statements described in the PFS are satisfied.

3.5 Step 3: Combine the Functional Areas to Form the System.

The objectives of this phase of the High Level Design development process are:

1. Find the common objects and assign ownership to these objects.

2. Find common data, create new objects to encapsulate this data, and assign ownership to these new objects.

3. Verify that ALL functions in the PFS are satisfied.

4. Validate the Internal system by verifying that specific manufacturing scenarios are satisfied.

3.5.1 Approach

Achieving the first two objectives will require a great deal of interaction among the entire application development team. A group workshop approach will used. A list of all objects and their functions will be compiled and used as a starting point to determine common objects and which functional areas use them. These common objects will be analysed, the requirements from the functional areas combined, and ownership assigned to complete the development of these objects.

Similarly, to determine the common data used by objects, a list of all data elements and the objects which use them will be compiled and used as a starting point. Again, we will use a group workshop approach. From this workshop, we would expect new objects to be created to encapsulate the common data (i.e., common data will be moved out of individual objects and moved 'up' to a larger object surrounding these objects). These new objects will then be assigned ownership, as the common objects.

To verify that all functions in the PFS are satisfied, the appropriate groups will work together and create object flows to validate the functions. This may also result in the creation of new objects.

The last objective, validating manufacturing scenarios, will involve a group creating object flows to verify that our objects will work together as a system.

3.5.2 Output for Step 3

The output for Step 3 will consist of a complete set of:

- Object Diagrams (with accompanying text description)

- Object Functional Diagrams (with accompanying text description)

- Object Flow Diagrams (with accompanying text and behaviour descriptions)

- Data Usage Tables

- Panel Usage Tables

- External API Tables

- Internal Interface Tables

All duplication will have been removed each object/operation will be unique. The combination of this data will constitute the system.

3.6 Step 5: Formal Documentation of Objects

The documentation will include a description of the object (stimuli, behaviour, and responses). The level of detail should include an understanding of the responses and the logic which produces these responses. This step will end when each object that is owned by the functional area has been fully documented,

3.6.1 Documentation Format

The documentation required is as follows:

1. ASSUMPTIONS

2. EXTERNAL VIEW (Object Diagram)

3. OVERVIEW OF OBJECT

> - Brief Description
> - List of API's and IPI's
> - State Data contained within the object

4. OPERATIONS

> - Narrative
> - Object Flow Diagram
> - Stimulus Response Matrix
> - Behaviour
> - API Charts
> - IPI Charts
> - Panel Charts

5. DATA USAGE CHART(S)

NOTE:-

1. Repeat Steps 3 and 4 for all objects that the function owns. Do not do this for objects defined as being common.

2. Design must be decomposed to the level where there are no calls to any other objects except either common or system ones.

3.6.2 Stimulus Response Matrix

This is to provide a verification on the Behaviour code and will show the stimulus (what's coming into the object operation), the current state (what condition the applicable state data is in), the response (what the output will be) and the new state (what condition the state data has been updated to, if applicable).

3.7 Step 6: HLD Formal Inspection

The purpose of the Inspection meeting is to identify errors NOT to find solutions.

When an error is found it is merely recorded, resolving the error is the responsibility of the author.

Inspection meetings should not last significantly longer than two hours as experience has shown that they become progressively less effective after this length of time. If the amount of work will clearly cause this guideline to be exceeded, two or more inspection meetings should be scheduled. In these situations each meeting should attempt to deal with a self-contained section of the material being reviewed.

4 CONCLUSIONS

The methodology described in this article was developed in parallel with the development of one portion of the program functional specification, the results were then applied to develop the other functional areas. This was due to object-oriented design being in its 'Industrial' infancy, this combined with the lack of any proven and fully documented methodology forced us to accept a high education overhead for the project.

Due to the lack of knowledge about object-oriented design development prior too this project we had difficulty in deciding what the initial system object definitions were, this was further increased when objects were designed with data and operations that enabled re-useability. Object-oriented design certainly has a place in the design of software systems. For a system of any complexity, one can use this method in a top down recursive approach (as described in this article). Object-oriented design has proved beneficial in the design of process systems.

The proposed method did not use an entity relationship diagram for the object identification, this useful tool missed some of the useful objects at the first level of recursion.

5 REFERENCES

Software Engineering - A Practitioner's approach, Second Edition, Pressman. 1983.

Principles of Information Systems Analysis and Design, H.D. Mills et al. Academic Press Inc. 1983.

SMALLTALK/V 286, DIGITALK Inc, L.A., CA90045

G. Booch, Software Engineering with Ada, Menlo Park CA. Benjamin/Cummings, 1983.

R. Abbott. Program design by Informal English Descriptions, Communications of the ACM, 26, Nov 1983

Chapter 12

Modelling, simulation and control of direct fired liquid gas vaporisers

A. Lees

1. INTRODUCTION

The British Gas Isle of Grain Storage Facility was built to meet the requirement for a large volume peak shave storage facility to serve the south east of England. Liquefied natural gas (LNG) was chosen as the most suitable means of providing this storage. The installation consists of four 21000 tonne LNG storage tanks, two liquefaction plants and a vaporisation/export system.

LNG is stored at a temperature of -160°C and a pressure of 100 mbar, and for export into the National Transmission System, the gas temperature must be raised to ambient and the pressure to grid conditions; between 40 and 70 bar.

Extraction from the tanks is performed by two sets of pumps capable of delivering 770 tonnes/hr of LNG at a maximum pressure of 70 bar. The vaporiser system consists of six direct fired heaters connected in parallel, each capable of vaporising 110 tonnes/hr of LNG to a minimum outlet temperature of 5°C. This makes the installation capable of supplying 780×10^6 sft^3/hr of natural gas to the National Transmission System.

A detailed description of the storage facility and its operating parameters can be found in reference 1.

2. VAPORISER CONTROL

Each vaporiser has two controllable inputs, fuel flow to the burners and LNG flow, which can both be modulated by control valves. A schematic of a vaporiser and its associated control valves is shown in Figure 1.

The original design for the vaporiser control system consisted of manual control of the LNG flow and automatic control of the fuel flow to achieve an outlet temperature of 5°C.

This semi-automatic design did not prove satisfactory in practice and resulted in the operation of the units being undertaken manually.

Since this was unnecessarily labour intensive, the site operators required a fully automatic control system.

The following control objectives were specified :-

i) full automatic temperature control over a LNG flow range of 25-110 tonnes/hr to achieve a stable outlet temperature of 5°C.

ii) The control system must withstand external effects such as rising and falling grid pressures and other plant e.g. starting/stopping pumps and other vaporisers. In particular, it must be able to withstand an instantaneous load increase or decrease of 25 tonnes/hr over the control range.

iii) automatic start-up of each vaporiser.

To develop a control strategy for the somewhat complex plant to meet these objectives, the site operators requested that a mathematical model be developed to enable the use of a simulation package to investigate and design a control system.

3. MODELLING AND IDENTIFICATION

Two possibilities existed for the modelling of the vaporiser. The first was to develop a model based on detailed knowledge of the physics of the plant which would accurately represent the plant at all operating conditions. However, this approach would have required a lot of test work and would have produced a complex model.

The second approach was to view the vaporiser as a 'black box' with two inputs, fuel flow and LNG flow and one output, gas outlet temperature. The aim would then be to find small signal dynamic models linking the output to the inputs. This would provide a simpler model but one which would only be valid over a limited operating range.

It was decided to follow the second approach as it was believed that this would provide a model relatively quickly and would be good enough to allow an investigation of control techniques.

The model structure using this approach is shown in figure 2. It shows a number of elements to be identified, in particular the dynamics associated with the vaporiser.

A number of tests were performed on-site to provide data for the model. These were as follows :-

i) steady state tests
ii) small amplitude step tests
iii) small amplitude sine wave tests

For each of the dynamic elements, small amplitude step tests (of approximately 1%) were performed to give approximate values of time constants, gain and time delay. Then small amplitude sine wave tests were performed to obtain gain and phase values which were plotted as Bode plots to enable derivation of transfer functions.

These tests were carried out at only two different LNG loads because of the long time constants and constraints on the export of LNG.

The parameters of the transfer functions identified are shown in Table 1.

In addition to the dynamic terms in the model, a number of steady state characteristics had to be identified. The fuel flow control valve was modelled using the Fisher Flow equation (ref 2).

The LNG flow control valve was modelled using the following equation for incompressible fluids :-

$$V = Cv\sqrt{dP}$$

V = Mass flow (tonnes/hr)
Cv=Flow coefficient (tonnes/hr/(lbf/in2)$^{\frac{1}{2}}$)
dP = Differential pressure (lbf/in2)

The pressure downstream of the fuel control valve was seen to be a function of fuel flow. The performance characteristic of the pumps meant that the pressure upstream of the LNG control valve reduced as the LNG flow increased. These characteristics were obtained and included in the model.

Finally a graph showing fuel flow against LNG flow in steady state was obtained as shown in figure 3.

4. MODEL VALIDATION AND CONTROL SYSTEM DESIGN

The data obtained was used to synthesise a model for solution on the British Gas Simulation package BSIMS. This is a suite of programs which facilitate the modelling of the elements identified and solves the differential equations involved. Graphical output enables the time response of system variables to be analysed.

The first objective was to validate the model. This was done by simulating two of the step tests performed during identification and comparing the results with those obtained on site. An example is shown in figure 4. It was considered that the model responses were close enough to be used in control system design work.

The next stage was the investigation of possible control schemes. The major problem in obtaining tight control was the presence of long time delays. Controllers must be detuned to ensure stability which degrades dynamic performance. Controllers were designed and tuned using classical linear design techniques; using Bode and Nichols plots to ensure adequate gain and phase margins. Simulation results were then used to confirm the values obtained and fine-tune if necessary.

A number of controllers were designed as illustrated in figure 5. A flow controller was designed for LNG flow to perform set point changes and maintain load under disturbance conditions. A proportional and integral (P+I) controller proved sufficient for this.

The fuel input was to be used to control the outlet temperature. With temperature feedback control alone, there was a delay between any change in LNG flow and the resulting temperature change. There was then a further delay before any compensation made to the fuel flow was reflected in the temperature.

Feedforward control was therefore included in an attempt to cancel out the effect of load disturbance dynamics. The identification results indicated that process dynamics varied with load so it was decided that to attempt to cancel out all temperature fluctuations due to load changes would be an unrealistic target. A constant gain feedforward controller based on the relationship of fuel flow to LNG flow as illustrated in figure 3 was used. A temperature feedback controller was then included to compensate for the inaccuracies in the feedforward controller.

The outputs of the feedforward loop and the temperature feedback loop were summed together to give the demanded fuel flow. To desensitise this to non-linearities in the control valve and pressure variations upstream, this was used to form the setpoint of a fuel flow control loop i.e. a cascade control system was used.

A number of other types of controller were investigated. Some incorporated phase advance to try to get over the problem of detuning. The use of a Smith Predictor controller was considered but this would have required a plant model valid over the total range of operation. It was considered that the control scheme as designed would be sufficiently robust and easy to implement with proprietary hardware and software.

5. CONTROL SYSTEM PERFORMANCE

The response of the system to three types of load change was examined to assess the suitability of the controllers designed. These were as follows :-

i) LNG flow set point change
ii) Pressure increase in supply header (vaporiser taken off- line)
iii) Pressure decrease in supply header (pump taken off-line)

In each case, the assumed initial LNG flow was 71.5 tonnes/hr and the temperature set point was 25°C. The results obtained are essentially small-signal results.

Figure 6 shows the LNG flow transient for a step change in flow set point to 80 tonnes/hr and figure 7 shows the resulting temperature transient. The temperature swings between 15°C and 35°C which is within the allowable limits.

When a vaporiser on the same supply header as the controlled one reduces its load, the effect is to increase the pressure in the header and hence increase the LNG flow through the vaporiser. This occurrence was simulated by instantaneously reducing the flow through a vaporiser to zero and recording transients for a vaporiser on the same header.

The resulting LNG flow transient is shown in figure 8 and the temperature transient in figure 9. Flow initially increased to 78.5 tonnes/hr as a result of a pressure increase of 5.5 bar in the header. This was brought back to set point within 3 minutes and the temperature deviation was less than 5°C in either direction.

When a pump supplying a header is switched off, the result is a reduction in pressure in the header and hence a reduction in LNG flow. This instance was simulated with a pump flow being instantaneously reduced to zero. An initial reduction of 5.2 tonnes/hr in LNG flow occurred, however the action of the LNG flow controller again compensated for this within 3 minutes. Again the temperature deviation was less than 5°C in either direction.

5.1 Comments

The above results indicate that the control system designed can meet the requirements specified earlier. The results for LNG flow set point changes show fairly large temperature deviations for the set point changes involved. Results from simulations of much larger set point changes indicate that the control scheme can cope with these if a rate limit is imposed on the set point changes. These results, however, infringe the small signal nature of the model.

6. IMPLEMENTATION OF THE CONTROL STRATEGY

The control strategy developed was implemented on a Negretti MPC84 multi-loop process controller.

As the simulation results were based on a small signal model, thought was given to the implication of operating over the whole vaporiser LNG flow range. Data gathered showed that the gains of process valves did not vary with load. However, the gains relating temperature changes to fuel and LNG flows did vary considerably. To overcome this it was decided to make the proportional gain of the temperature controller vary with LNG flow, i.e. adaptive gain. Theoretical calculations of the gains indicated that both were inversely proportional to LNG flow. In order to keep the overall process loop gain constant, therefore, the adaptive gain was made proportional to LNG flow.

The control scheme used was that shown in figure 5.

6.1 On-site Validation Of Control Strategy

A number of tests were performed on a vaporiser using the control scheme as implemented on the Negretti process controller. They were performed while other vaporisers were also operating; hence all the problems of vaporiser interaction were present.

Figure 10 shows the response when the LNG set point was increased from 77 tonnes/hr to 100 tonnes/hr. As a consequence of a rate limit of 2.5 tonnes/hr/minute on the set point, the LNG flow rate took 11.5 minutes to reach its final value. The minimum value of the outlet gas temperature was 13.2°C.

As a consequence of load changes on other vaporisers, a number of pressure changes occurred in the header, causing the LNG flow changes of up to 15 tonnes/hr that can be seen. These were compensated for by the LNG flow controller in 15-20 seconds. Figure 11 shows a transient of this type in more detail; in particular the responses of the LNG flow controller, the fuel flow controller and feedforward controller can be seen.

Tests were also performed where pumps and other vaporisers were switched off.

The final test was of automatic start-up. This could not be reasonably be simulated using the model previously described. An algorithm was written and implemented for start-up. The LNG set point was driven up to a base load - 40 tonnes/hr was used in this case.

The feedforward controller then brought the fuel flow up to match this flow. No further LNG flow set point changes were possible for 10 minutes to allow the temperature to stabilise; the temperature controller switching in as the temperature came within its control range.

7. CONCLUSIONS

A classical exercise in modelling and simulation has enabled the design and implementation of a control scheme for the vaporisers at the British Gas Isle of Grain Storage Facility to be achieved. A small signal model was developed from on-site plant data and validated.

A control scheme was then developed using the derived model which simulations indicated would meet the control specification. This work gave sufficient confidence in the control scheme for implementation to go ahead and plant trials to be performed.

The control scheme was implemented with small modifications including adaptive gain and autostart and a number of successful plant trials were performed, meeting the original specification for automatic control.

8. ACKNOWLEDGEMENTS

The author wishes to thank British Gas for permission to publish this paper and colleagues within British Gas who have provided valuable assistance and advice.
This paper is based on a paper previously presented at an I.Mech.E. seminar 'Simulation In The Process Industries' on April 9 1986.

9. REFERENCES

1. The Isle of Grain LNG Installation. A.J.Findlay and G.W.Spicer Gas Engineering and Management. February 1983.

2. Development of a Universal Gas Sizing Equation for Control Valves. James F.Buresh and Charles B.Schuder. Fisher Governor Company. ISA Transactions - Vol.3, No.4, October 1964.

FORM OF SYNTHESIZED TRANSFER FUNCTION

$$\frac{K \ W_n^2}{S^2 + 2 \ \zeta \ W_n \ S + W_n^2} \ e^{- ST_d}$$

K = dc gain
ζ = damping ratio
W_n = undamped natural frequency
T_d = time delay
S = Laplace operator

TRANSFER FUNCTION	LNG FLOW TONNES/HR	ζ	Wn (mHz)	Td (SEC)	K	
					FROM DYNAMIC TEST	FROM STEP TEST
					($°C/(sm^3/hr)$)	
Fuel Flow/ Gas Temperature	71 85	1.01 0.96	2.0 2.2	40 40	0.115 0.082	0.123 0.068
					($°C/(tonne/hr)$)	
LNG Flow/ Gas Temperature	71 86	0.49 –	3.5 –	30 30	-1.48 –	-1.93 -1.51
					(No units)	
Fuel Valve Control Signal/ Fuel Valve Position	–	1.03	160	0.5	1.0	1.0
LNG Valve Control Signal/ LNG Valve Position	–	0.28	81	1.0	1.0	1.0

TABLE 1

TRANSFER FUNCTION PARAMETERS

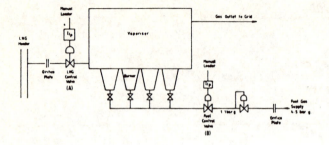

FIG.1   SCHEMATIC OF VAPORISER AND CONTROL VALVES

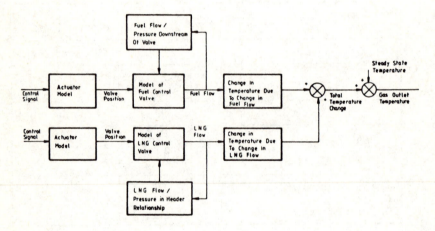

FIG.2   MODEL STRUCTURE FOR VAPORISER

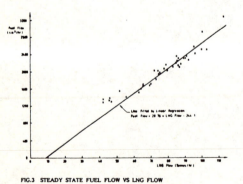

FIG.3   STEADY STATE FUEL FLOW VS LNG FLOW

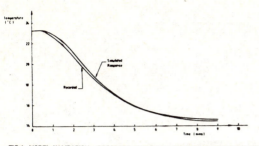

FIG.4 MODEL VALIDATION - RECORDED AND SIMULATED TEMPERATURE RESPONSES TO FUEL FLOW STEP CHANGE

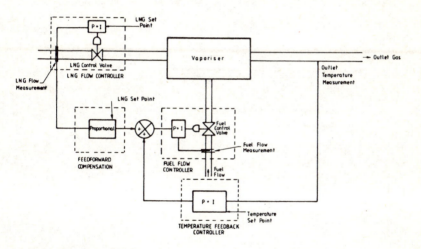

FIG.5 SCHEMATIC OF CONTROL SCHEME

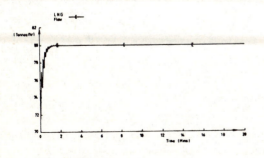

FIG.6 LNG SET POINT STEP CHANGE TO 80 TONNES/HR

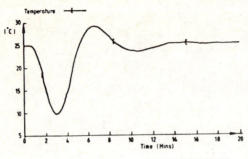

FIG.7 TEMPERATURE RESPONSE TO LNG SET POINT STEP CHANGE

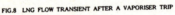

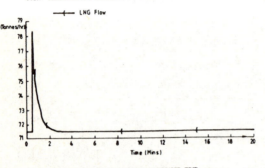

FIG.8 LNG FLOW TRANSIENT AFTER A VAPORISER TRIP

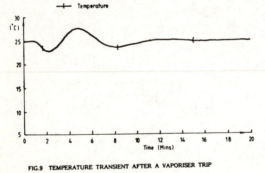

FIG.9 TEMPERATURE TRANSIENT AFTER A VAPORISER TRIP

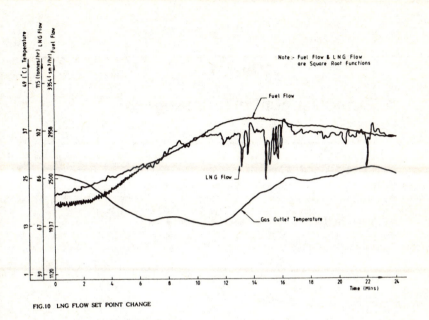

FIG.10 LNG FLOW SET POINT CHANGE

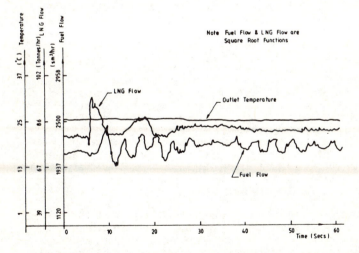

FIG.11 FUEL FLOW, LNG FLOW AND FEED FORWARD CONTROLLER RESPONSES TO PRESSURE INCREASE IN HEADER

Chapter 13

Computer control for patient care

D. A. Linkens

An overview is given of a number of computer controlled drug administration schemes in clinical medicine. These range from systems which have
been used on many patients in intensive care to experimental schemes which
require further clinical evaluation. The control algorithms used vary
from simple PI controllers to multi-mode adaptive techniques. Measurement of relevant clinical variables is often a major problem, and the use
of extended Kalman filtering is described for the estimation of unmeasur-
able states. Recent developments in expert control are also described.

1. INTRODUCTION

It is commonly recognised that the human body contains a large number of
control mechanisms with complex behavioural characteristics and interconnections. It is not surprising, therefore, that the techniques developed in the analysis of control systems have been applied to many aspects of patient care. In particular, techniques of systems identification are being applied to a wide range of biomedical sub-systems to elu-
cidate the dynamic characteristics of component parts of the human physiological system. The aim of this chapter, however, it not to cover this
aspect of systems analysis in biomedicine, but to describe current research into the use of external automatic controllers to regulate certain
variables in the body. Thus, instead of inherent homeostasis mechanisms,
we shall consider the design and analysis of control systems where the
body is only part (albeit the main part) within the loop. External control systems become necessary only under conditions of illness, and in
the following sections we shall consider both the direct control of important physiological variables such as blood pressure and blood glucose
levels, and clinical assistance under operating theatre conditions involving such things as levels of unconsciousness and relaxation. Although all the applications described involve a drug infusion regime as
the output of the automatic controller, it should be noted that any other
physical quantity which would affect the controlled variable could be considered eg electrical impulse regime.

An important requirement for an infusion controller is that the control
system interfaces in a 'user friendly' way with the clinician or nurse
who supervises its operation. It must also alert the ward staff to
sudden changes in the patient's condition or component failures in the
controlled infusion system. Alarms are needed to indicate emergency
situations such as cardiac arrest (low blood pressure), nurse/patient
noise (taking blood samples), maximum/minimum dose warnings, pump mal-

function, pump/computer communication failure and computer peripheral faults. A visual display unit provides excellent facilities for interfacing the system with the staff of the ward and for providing alarm warnings.

Tardiness in applying computer control techniques in medicine is related to the complex nature of human physiological response to controlled infusions, the difficulty in quantifying those responses and the stringent requirement for patient safety. The major difficulties facing the control system designer in the medical environment can best be illustrated by looking at some specific applications of closed-loop computer control to patient care.

2. POST-OPERATIVE BLOOD PRESSURE CONTROL

In cardiac surgical and intensive care wards the manual management of blood pressure is a routine task. However, in these wards there are many other demands on the time of the staff. This may result in adjustments to vasoactive drug infusion levels only being possible at 15-30 minute intervals. In the case of patients with highly unstable autoregulatory systems, adjustments at such widely spaced intervals may be quite inadequate to successfully maintain the patient's blood pressure within the required target zone. In both these situations, if the medical or nursing staff can be relieved of the routine control tasks, they would be better able to give their attention to more essential aspects of patient care where human perception and understanding are vital in avoiding crisis and encouraging recovery.

Large numbers of patients have been treated post-operatively using automated feedback control of blood-pressure. Over the past few years more than 10,000 hypertensive patients have been treated in this way by L C Sheppard and his co-workers at the University of Alabama, Birmingham, USA. Techniques involving simple classical control, systems identification and adaptive control strategies have all been applied to this situation, and clearly illustrate the way in which further knowledge about the dynamics of a physiological system and improved control can be achieved via an inter-disciplinary research and development effort.

Before attempting closed-loop control some knowledge of systems dynamics is essential, and this has been performed using PRBS excitation for the blood pressure system (Sheppard and Sayers, 1977). In this work, blood-pressure measurement was taken as the mean arterial pressure (MAP), and its response to a number of hypotensive agents was determined. The later computer-controlled work has concentrated on the use of one of these drugs, sodium nitroprusside. Cross-correlation between the MAP response and the PRBS drug input revealed information about the impulse response of the physiological system. This indicated the presence of a time delay of about a single blood circulation time (30-40 sec) together with at least one time constant. The impulse response typically showed a one-minute period to the first minimum, 3-minutes period to second minimum, and a return to baseline within 5 minutes. Detailed modelling of these dynamic responses was performed later, but the initial information was sufficient for the design of simple classical PI feedback control.

Using typical impulse responses, simulation studies were used to select

suitable parameters for a PID controller.   It was found that these
values required 'tuning' when closed-loop control was implemented
(Sheppard et al, 1979).   Although this was partly due to simplifica-
tions in the modelling, it was also due to patient-to-patient variability in system parameters.   To give satisfactory response it was found
to be necessary to include a 'decision table' in cascade with the PI
controller to limit the drug input within a number of clinical and physio
logical constraints.   In this way, the control system became non-linear
in behaviour.   An example of the improvement in the blood-pressure variability caused by using automatic rather than manual control is shown
in Fig 1.   In a range of trials, significant improvement in the quality
of control was achieved with the use of this constrained PI regulator.

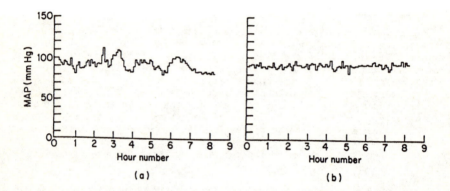

Fig 1   Blood pressure during infusion of sodium nitroprusside under
        a) manual control b) computer control (from Sheppard et al,
        1979)

More detailed modelling studies have revealed further dynamic features
in the blood-pressure response, including a recirculation effect and
background activity containing both stochastic and sinusoidal components
(Slate et al, 1979).   The detailed model obtained as a result of these
studies is shown in Fig 2.   The transfer function $G_d(s)$ contains two
time delays and one exponential time constant, and is given by

$$G_d(s) = \frac{Ke^{-T_i s}(1 + \alpha e^{-T_c s})}{(1 + \tau s)}$$ (1)

where T_i (30 seconds) represents the initial transport-lag for the
drug from the injection site, and T_c (45 seconds) represents the recirculation time. α is the fraction of a drug recirculating, and τ is
the time constant associated with drug metabolism.   Use of this model
for a range of drug sensitivities K under the fixed controller strategy
previously mentioned showed that such a structure could be unstable,
and this has been confirmed in clinical usage.   Because of the non-

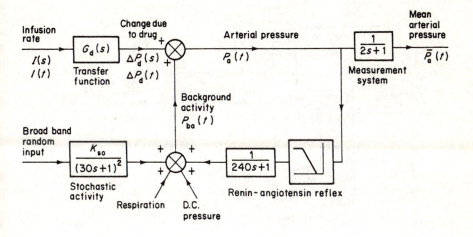

Fig 2   Blood pressure model for vasoactive drug infusion
(from Slate et al, 1979)

robust nature of this method of control, an adaptive control structure
has been investigated.

The adaptive control scheme studied by Slate (1980) is shown diagramma-
tically in Fig 3.   This scheme contains a number of control modes, and
multirate sampling and filtering of the measured signals.   Thus, blood
pressure is sampled with a period of 1 second.   This noisy signal is
low-pass filtered and subsequently re-sampled at a 2 second period.
This signal is used to form the closed-loop error signal which is further
sampled at a 10-second period for processing the controlled algorithm.
The derivative of the error is also required, and this is obtained via
the 2-second sampled signal by passing it through a 3-point differentiator (which attenuates high frequency noise), followed by a low-pass filter, with subsequent sampling at the 10 second period.   Based upon these
error and derivative of error signals, the coordinator selects either a
transient control mode or a regulator mode to calculate the incremental
action.   The coordinator actions can be represented by a phase-plane
diagram of error versus error rate.   In the transient control mode, the
increment is calculated by a relay-type controller that includes a
Smith time-delay compensator.   A gain scheduler is included in the
transient controller to produce a fast speed of response to hypotensive
pressure transients.   In the regulator mode, the infusion rate increment
is calculated from the error and its derivative.   Adaptation to the
patient drug gain is accomplished in an initial period after closed-loop
control has begun.   A recursive least-squares estimation algorithm is
used to provide a measure of the drug gain parameter from data acquired
at 2 second intervals.   The estimated patient gain and the variance of
this estimate are used to adjust (at 10 second intervals) the overall
gain of the regulator, the relay output levels, and the Smith compensa-

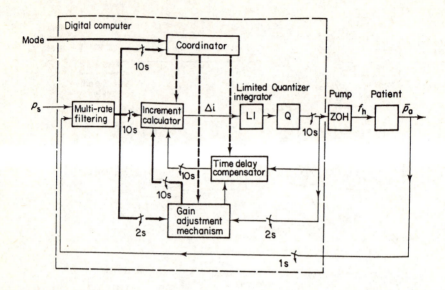

Fig 3   A self-adaptive control strategy for blood-pressure control
of hypertensive patients (from Slate, 1980)

tor gains.   These adjustments are managed by decisions made by the coordinator.   Although the performance of this multirate adaptive scheme
was assessed only in terms of the model described   above, it has subsequently been implemented in microcomputer format and evaluated in dog
trials.

A number of other control designs have been considered using the Slate
model.   These include adaptive gain control (Koivo, 1980) and a sub-
optimal self-tuning design (Walker et al, 1982) based on the well-known
Clarke and Gawthrop algorithm.   This approach has also been used by
Millard et al (1988) in the control of blood pressure during surgery.
A range of clinical trials have been performed, including the use of
phenylephrine during lower abdominal surgery, isoflurane for induced
hypotension during ENT surgery, and SNP as used by Sheppard and co-
workers.

An alternative pole-placement algorithm for self-tuning has been investigated via extensive simulation by Mansour and Linkens (1988).   A range
of effects was considered including noise, respiration artefact, switched
reflex loops due to renin-angiotensin and epinephrine, drug saturation,
and time-varying parameters.   Satisfactory performance was maintained
provided the system was 'jacketted' to prevent loss of good identification during nonlinear regions of operation.   The Slate model includes a
time-delay, and a series of studies were performed incorporating the con-

cept of Smith prediction into the self-tuning regime (Mansour and Linkens, 1989). This technique gave good results, including situations involving mismatch between the 'patient' and the model assumed in the Smith predictor. Related work has been performed by Packer et al (1988) using dopamine for seriously ill hypotensive patients. Their approach is to continuously test patient sensitivity during the normal operation of the controller. The response to step changes in the infusion rate which occur at the implementation of each control decision can be used for this form of adaptation although noise in the MAP signal may lead to erroneous estimations of sensitivity. This can be minimised by averaging over a number of estimations.

The above work considers the infusion of one drug only, but in clinical practice blood pressure is frequently controlled using multiple drug infusions. In recent years, several multivariable drug delivery systems have been investigated. Serna et al (1983) explored the simultaneous control of cardiac output (CO) and the mean arterial pressure (MAP) with sodium introprusside (SNP) and dopamine (DOP) using model reference adaptive control. McInnis et al (1985) presented a scheme for controlling arterial and venous pressure via inotropic and vasoactive drug infusion. They used an optimal one-step-ahead controller and a bilinear dynamic model for their studies.

Gregory et al (1987) simultaneously controlled MAP and CO in anaesthetized dogs using infusion of SNP and dobutamine. The self-adaptive design was based on a control advanced moving average controller (CAMAC), which determines the controller outputs at every sample to minimise the difference between the set point and the predicted output at a time which is equal to or greater than the system's time delay. In more detailed modelling studies by Linkens and Mansour (1989), the simultaneous control of MAP and CO via drug infusions of SNP and DOP was investigated. In this case, non-pulsatile models of the cardiovascular system due to Moller (1983) and Wesseling (1982) were employed. The control algorithm was a multivariable self-tuning pole-placement version due to Sirisena and Teng (1986) which has advantages in its implementation simplicity. The system has been shown to be capable of dealing satisfactorily with non-linearities in the models and limiting effects of the infusion pump drive.

3. ANAESTHESIA CONTROL

Three major areas of responsibility for the anaesthetist are those of unconsciousness in the operating theatre, drug-induced muscle paralysis during operations, and pain relief (analgaesia) during post-operative care. Each of these areas have been explored with respect to the use of feedback control and are reviewed in the following sections.

3.1 Drug-induced Unconsciousness

This is the most commonly recognised role of the anaesthetist, and reveals the earliest attempts to produce automated feedback control in this discipline, under the title of the 'servo-anaesthetiser' (Bickford, 1949). As pointed out in a review article by Chilcoat (1980) the major problem with this, and subsequent attempts, was the dubious nature of the use of the EEG as a measurement for the depth of unconsciousness. The special problems which provide difficulties in this field for the system designer mainly relate to patient variability and accurate measurement of physiological parameters:

(a) Patients vary widely in their response to anaesthetics. Sensitivi-
    ty to a particular anaesthetic agent varies from patient to patient
    and, for a given patient, may vary with time and with the level of
    drug administration. The controller must monitor and quantify
    changes in sensitivity and adapt the control strategy to the new
    sensitivity conditions.

(b) A reliable method must be used to determine and quantify the depth
    of anaesthesia. The controlled variables should reflect the depth
    of anaesthesia rather than the level of the drug in the patient.
    The output of the measurement systems will include noise generated
    by non-physiological artefacts.

(c) It is important that potentially critical trends in patient condi-
    tion be detected so that appropriate preventive action can be taken.
    The controller must provide well designed data display and manual
    intervention facilities for the anaesthetist who provides overall
    supervisory control and must take the responsibility for averting any
    crisis.

(d) Since anaesthetics affect many physiological systems, both desirable
    and undesirable effects must be balanced. The controller should be
    multivariable in that it must measure and control a number of rele-
    vant parameters simultaneously.

A number of variables have been used as measures of the level of anaesthesia. These include electroencephalograph readings (EEG) for thio-
pental, ether, halogen and methohexital; mean arterial blood pressure
(MAP) for halogen and evoked muscle action potential. Where multivariable control has been implemented the additional variables monitored
and controlled have included inspired oxygen concentration, end-tidal
partial pressure of carbon dioxide, and anaesthetic agent concentration
at the end of expiration and breathing circuit volume. While some
research groups have been successful at partial automation and anaesthesia, a comprehensive multivariable system for routine use in the
operating theatre is not yet available.

In the work of Coles et al (1973) a multivariable system was proposed
to control several inter-related variables in sheep, including arterial
blood pressure, as a measure of depth of anaesthesia. Their system
maintained constant levels of end-expired CO_2 and inspired O_2, plus the
reservoir-bag volume in a closed anaesthetic breathing system. Oscillation of the controlled variable proved to be a problem in the blood
pressure loop, a point also noticed by Smith and Schwede (1972), illustrating the problems of good feedback control in the presence of time
delays, as mentioned in the previous section.

The measurement of unconsciousness has also been approached using
'clinical signs', such as pulse rate, arterial blood pressure, respiratory rate, tidal volume, sweating movement etc. In the work by Suppan
(1972), one of these variables could be selected by the anaesthetist
for control via a motor-driven vapouriser administering halothane.
In later work a 'clinical scoring' approach has been used, whereby a
number of clinical signs are summed together with selected weighting to

give a single measure of depth of anaesthesia (Davies et al, 1982). This has been used clinically with a micro-computer based system which allows for entry of the scores via a keyboard, and provides control of a motor-driven syringe for the administration of drugs intravenously.

The problem of designing adequate feedback controllers for biological systems, which often contain time delay components, is compounded by the large patient-to-patient variations in parameters for the dynamic models. This has prompted the investigation of adaptive control schemes for depth of anaesthesia. Beneken et al (1979) showed via a multicompartment model of the anaesthetic system that a parameter-estimation adaptive control system could give superior performance to a conventional fixed parameter controller. This type of approach has been followed by Tatnall and Morris (1977) on neonates. Using a twelve compartment body model, they showed via simulation that a fixed PI controller would not give acceptable control when allowances are made for differences in patient uptake characteristics. Successful clinical trials have been undertaken using their structure of self-adaptive control (Tatnall et al, 1981). It should be noted that in this work depth of unconsciousness is being inferred solely from levels of gas concentration in the alveoli.

An alternative approach to the feedback methods described above is to use open-loop control based on a known model of drug uptake. In this case the dynamic model is used to predict the time-course drug level in brain tissue, which is presumed to determine the depth of anaesthesia. Purely open-loop control of this nature would be unworkable because of variations in patient dynamics, and hence a number of schemes have been attempted in which the model is updated based on infrequent measurement of certain process variables. Thus, Mapleson et al (1974) reported trials on dogs in which they attempted to achieve an arterial halothane tension of 4mm Hg within 5 min, and maintenance of this tension for a further 75 min. An extensive model was used, whose parameters were updated every 10 minutes, based on measurements of cardiac output and alveolar ventilation. A more extensive system has been reported by Mapleson et al (1980) which automatically controls the set-point of a vaporiser. It does this based on a computed brain tension of anaesthetic agent, obtained from a detailed model of uptake and distribution. The model is initially set up based on standard values and body mass, and subsequently updated every 10 min using measurements of cardiac output, alveolar ventilation and arterial blood tension. The anaesthetist is included in the loop by means of manual control of the N_2/O_2 mixture which carried the volatile agent. He specifies the desired brain level of the volatiles in terms which include the contribution of N_2O.

3.2 Drug-induced Muscle Relaxation

In certain operations, such as abdominal surgery, correct levels of patient muscle relaxation are necessary. A number of drugs exist for this purpose, and the type referred to in this section are of the non-depolarising form, such as pancuronium. To attain automated muscle relaxation, a suitable measurement variable must be defined. The level of muscle relaxation can be simply quantified either in terms of an evoked electromyograph (EMG) or an evoked tension response. In the work reported here, the former method has been adopted using supramaximal stimulation at a frequency of 0.1Hz at the ulnar nerve above the

elbow.   The resulting EMG is measured using surface electrodes taped
to the hand.   The EMG signals are amplified with a gain of 1000 at a
bandwidth of 8Hz to 10KHz, then rectified, integrated and finally stored
in a sample-hold amplifier.

Computer control of muscle relaxation using stimulation of the masseter
muscle on the face of sheep has been achieved, (Cass et al, 1976).   In
human trials using a simple proportional gain feedback controller, satisfactory regulation was achieved with a mean level of 74% paralysis for
an 80% set-point (Brown et al, 1980).   This offset has been successfully removed using a PI controller whose parameters were set using the
Ziegler-Nichols tuning method.   Using a fixed PI controller occasionally
gives an oscillatory closed-loop response, as already noted in the section on blood-pressure control.   This emphasises the need to identify
the dose-response model for muscle paralysis.

The model required here is a combination of the drug pharmacokinetics
(ie drug-dose to blood level concentration) and pharmacodynamics (ie
blood level concentration to evoked EMG response).   Instead of classical bolus injection methods for pharmacokinetic determination, the use
of PRBS excitation has been made for model identification (Linkens et
al, 1981).   Using bit intervals of 33.3 sec or 100 sec with a sequence
length of 63, successful identification has been made in dog trials.
Employing an off-line technique, identification of the data has been
obtained using a generalised least squares package.   This revealed the
presence of a pure time delay with mean value of 64 sec and two exponential time constants with mean values of 2.7 min and 20.1 min.   These
parameters had a range of about 4:1 in a small number of trials, illustrating the large variability in dynamics which is common in biological
systems.

The feedback system described is being used in a range of microcomputer
control studies.   These include attempts to quantify the effects of
other drugs, such as tranquilisers, on relaxation levels (Asbury et al,
1981), and the interacting effects of other anaesthetic agents such as
halothane and ethrane in potentiating paralysis (Asbury et al, 1982).
Simulation studies have shown the desirability of using Smith predictor
schemes for counteraction of the pure time delay, based on a sensitivity
analysis using the parameter ranges identified previously (Linkens et
al, 1982).   Adaptive control schemes have also been investigated using
a pole-assignment form of self-tuning controller, with successful clinical trials (Linkens et al, 1982).   The identified model for relaxant
dynamics is suitable for the application of a self-tuning PID control
algorithm due to Keviczky.   This has been studied for the non-linear
model and developed successfully via simulation studies and implemented
clinically (Linkens and Denai, 1989).

On-line recursive identification algorithms have been implemented and
tried successfully in dog studies under either open-loop or closed-loop
conditions.   The aim, therefore, is to use a form of adaptive control
which will give simultaneous control and identification of drug dynamics.   This latter aspect is of interest since some of the newer relaxant drugs, such as vecuronium, do not have easy methods for determination of drug pharmacokinetics via classical bolus injection methods.
Thus, it can be seen that as well as automated regulation of relaxation
levels, studies of this nature can give more detailed quantification of
interacting drug effects.

3.3 Post-operative Pain Relief

The control of pain following major surgery, such as total hip replacement, is an important clinical requirement.   In conventional clinical
practice the necessary feedback is provided by nursing staff who admini-
ster pain-relief drugs based on observation of the patient at about four
hour intervals.   Such slow sampling cannot give optimal relief of pain.
In an automatic 'demand analgesia' system, feedback is provided by the
patient who is equipped with a button and instructed to press it whenever
he feels uncomfortable.

The use of a simple proportional gain controller in such a system is
described by Jacobs et al (1981).   In this work, pain is quantified as
the rate of button-pushing by the patient.   Proportional control in
this case is non-optimal because it requires the patient to experience
some pain before he can receive any pain-relief drug.   This non-optima-
lity cannot be removed simply in this case by introducing integral control action because of the absence of negative demands.   One commercial
system attempts to overcome this severe output nonlinearity by introducing what amounts to a positive non-zero desired value of pain, thus
allowing integral action control (White et al, 1979).

In work by Jacobs and his co-workers (Jacobs et al, 1982, Reasbeck, 1982)
the nonlinear problem is overcome, with a claimed significant improvement
in performance, by using a separated stochastic control in which a non-
linear state estimation is cascaded with a one-step-ahead control law.
The nonlinear estimator comprises a Bayes algorithm cascaded with an extended Kalman filter.   The model used in this work is shown in Fig 4,
where perceived pain y is assumed to depend on the difference between
'comfort' due to drug administration and 'discomfort' due to surgery.
The pharmacokinetic relationship between drug infusion rate u and brain
tissue drug concentration is assumed to be triexponential.

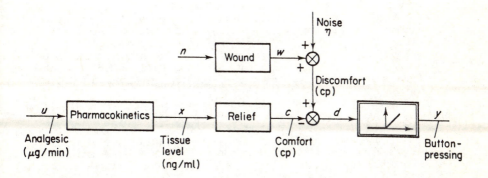

Fig 4   Structure of a mathematical model for control of post-
operative pain (from Jacobs et al, 1982)

Little is known about the relationship between comfort and tissue drug
level and here it is assumed to be a simple constant of proportionality,
whose estimation is itself of importance, and is included in the Kalman
filter.   The discomfort is modelled as the sum of an exponentially de-
caying stochastic term w, which represents a healing wound, plus white
noise η.   Two nonlinearities arise in the relationship between the
states x and the output y.   One is in the estimation specifying nett
pain d, and the other is the severe demand nonlinearity y(d) in Fig 4.
Whenever d is greater than zero, the estimation of states can be done
via standard extended Kalman filter (Jazwinski, 1970).   The control law
was designed to make the comfort c greater than the predictable component w of discomfort by an amount proportional to the magnitude of the
unpredictable component η of discomfort.

Simulation studies have been undertaken (Reasbeck, 1982) of the three
systems referred to above, from which it was shown that the Kalman filter approach gave considerably fewer analgesia demands than the simple
proportional or commercial controllers (23 demands against 166 or 122).
This improvement was achieved, however, with a higher total drug consumption (674 μg against 332 μg or 473 μg).   It is suggested that good
estimates of the relief can be obtained using this method of stochastic
control.

4.   CONTROL OF BLOOD GLUCOSE LEVELS

Diabetes is characterised by a chronically high blood glucose concentration.   There are two main clinical categories of diabetes: type 1
termed juvenile onset, and type 2 termed maturity onset.   In both
types the basic defect is thought to be a relative or absolute deficiency of insulin in the blood, the glucose lowering hormone produced
by the pancreas.   Type 1 diabetics have the greatest deficiency of insulin and are normally treated with one or more daily injections of insulin.

The use of bolus subcutaneous injections of insulin in the control of
blood glucose concentration levels in a diabetic patient is normal clinical practice.   Better physiological response should be obtained if
the glucose level is monitored regularly and insulin delivered in a
regime more closely resembling the normal release mechanisms.   In
simple terms, the B-cell delivers insulin into the blood stream at two
rates: a continuous, slow basal rate, which controls glucose output from
the liver, and mealtime bursts which dispose of the digested nutrients.

With the advent of continuous blood monitoring and frequent sampling
profiles it has become apparent that in most insulin-dependent diabetic
patients it is difficult to sustain near normal blood glucose concentrations.   The so-called 'artificial pancreas' is a closed-loop system,
ie an extra-corporeal blood glucose sensor is coupled to a computer
which controls the rate of infusion of insulin into a peripheral vein
so as to maintain normoglycaemia (Albisser and Leibel, 1977).   Although very successful in maintaining normoglycaemia in diabetic
patients for up to a few days, it has major disadvantages for long term
use.   Thus, it is limited by its bulk, complexity and cost, and its
use of the intravenous route for insulin administration and blood sampling.   Prolonged infusions carry the risk of thrombosis and infection.

The above considerations have prompted simpler portable infusion pumps
without glucose sensing, and thus operate under open-loop conditions.
In this approach, continuous subcutaneous insulin infusion (CSII) employs
a portable electromechanical syringe pump capable of delivering insulin
at two fixed rates. The lower level is continuous, while the higher
rate is electrically engaged by the patient 30 minutes before each main
meal (Pickup and Keen, 1980). A number of studies have now shown that
near physiological blood glucose concentrations can be maintained by CSII
for a number of days in ambulatory diabetics supervised in hospital.
In terms of long-period out-patient treatment a group of 6 diabetics
treated at home via CSII for 2-4 months achieved overall mean blood glucose values varying from 4.8 to 7.7 mmol/1 (Pickup et al, 1979). There
are many intermediary metabolites and hormones which have abnormal concentrations in diabetics, and may be contributory factors in the patho-
logical processes. Treatment via CSII has been associated with a return to near-normal blood levels of lactate, pyrovate, 3-hydroxybutyrate,
alanine, cholesterol, triglyceride, and free fatty acids.

In spite of the success of CSII open-loop control in certain cases, there
remain 'brittle' diabetics in whom large, fast and unpredictable swings
in blood glucose occur and for which open-loop control cannot help. In
such cases feedback control is the only way in which near-normal blood
glucose levels could be achieved. Similarly, it has been observed that
CSII does not allow for classical output disturbances caused by such
things as severe stress or intermittent illness. The comment has been
made that CSII should only be used therefore under close medical supervision - ie under closed-loop control! The situation begins to approximate to the combined open-loop/closed-loop system described for drug-
induced unconsciousness.

A number of control algorithms have been proposed for closed-loop control 'artificial pancreas'. The simplest one was an on-off mechanism
which activated syringe pumps containing insulin and glucagon depending
on threshold levels of blood glucose (Kadish, 1964). Kline et al (1968)
used a simple proportional controller driven by the deviation from normoglycaemia. In the work by Albisser three relationships for controller
synthesis were used. Two of these are sigmoidal shaped curves relating insulin and glucagon delivery rates to the measured level of gly-
caemia, while the third is a difference factor which depends on the rate
of change of blood glucose. These relationships are summarised by

$$\text{Insulin delivery} = 200 \left(1 + \tanh(G_p - 140)/25\right)$$

$$\text{Glucagon delivery} = 1.5 \left(1 - \tanh(G - 50)/70\right)$$

$$G_p = G + (A^3 + 10A) \tag{2}$$

where

$$G = \text{glycaemia}$$

$$G_p = \text{projected glycaemia}$$

$$A = \text{rate of change of glycaemia.}$$

An adaptive control scheme which allows for patient parameter variations
and selectable blood glucose profiles has been studied on dogs by Kondo

et al (1982).   Their adaptive scheme is outlined in Fig 5.   The reference model used is a second order linearised model similar to that of Ackerman et al (1964) and given by

$$\dot{x}_1 = -p_1(x_1-x_f) - p_2x_2 + p_3v$$
$$\dot{x}_2 = -p_4x_2 + p_5x_1 + u$$

(3)

where x_1 is blood glucose, x_f is its fasting value, and x_2 is blood insulin level.   Controller 1 is designed to minimise the cost function

$$J_r = \Sigma(y_r^2(k) + \rho_r u_r^2(k))$$

(4)

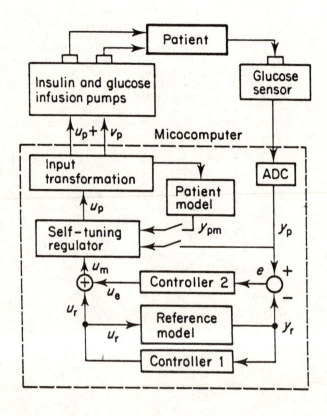

Fig 5   Schematic diagram of an adaptive and optimal blood glucose control system (from Kondo et al, 1982)

where proper choice of ρ_r gives the desired clinical profile. Controller 2 is designed from optimal principles to minimise the output error, $y_p - y_r$, which is caused by errors in initial estimates of parameters and disturbances due to meals. The self-tuning regulator is used to make the input/output relationship of the diabetic subject the same as that of the reference model. To do this required recursive estimation of the patient parameters, together with updating of the controller law. The scheme has been successfully used in simulation studies including nonlinear patient dynamics, and in dog trials (Sano, 1986).

The main disadvantages of these closed-loop devices are the large size and complexity and the intravenous route of insulin delivery. The latter limits the duration of treatment, because of the risks of thrombosis and infection. The bulk restricts the device to virtually bedside use and also prevents long-term experiments. Miniaturisation has been partly hampered by the lack of a suitable implantable glucose sensor. There is no shortage of feasible sensing strategies but some of the problems which must be solved are biocompatibility, sensor drift in vivo, storage and shelf life, cost, ease of manufacture and, of course, complete reliability.

5. CONTROL OF RESPIRATORY VARIABLES

The goal of the respiratory system, working together with the cardiovascular system, is to supply sufficient oxygen and to remove sufficient carbon dioxide from the tissues. To perform this task, a single variable cannot, in general, be used to describe the performance of the system. In the automatic control of artificial ventilators a number of quantities have been measured. These include partial pressure of arterial $CO_2 (P_aCO_2)$, partial pressure of alveolar $CO_2 (P_ACO_2)$, arterial $O_2 (P_aO_2)$, pH, and CO_2 production (VCO_2). The controlled variables similarly show a wide range including minute ventilation (adjusted either via the respiratory rate or the tidal volume), fraction of CO_2 in inspired gas $(fICO_2)$, and fraction of O_2 in inspired gas (fIO_2).

Under normal physiological conditions P_aCO_2 is kept substantially constant, and hence the aim of some automatic control systems has been to keep this constant. P_aCO_2 is not, however, easy to measure, and hence end-tidal CO_2 which is obtainable non-invasively has commonly been used. In normal lungs end-tidal CO_2 approximates to P_ACO_2, which in turn approximates to P_aCO_2. An early clinical example of this approach was by Frumin (1956), while microprocessor control has been demonstrated by Westonkow et al (1980). Related work reported more recently is that of Bhansali and Rowley (1984), Chapman et al (1985) and Giard et al (1985). An alternative approach has been to use a pH sensor and to maintain arterial pH constant by adjustment of the tidal volume (Coon et al, 1978), and related work is that of East et al (1982). Some systems have been designed to adjust the inspired fraction of O_2 in addition to that of CO_2 eg Chambille et al (1975).

The control of ventilatory waveform has also been considered, especially for positive pressure artificial ventilators. These force air into the lungs in a manner which is different from the normal negative pressure spontaneous breathing condition.   Adverse effects caused by positive pressure ventilation are considered to be correlated with the average alveolar pressure, and can be influenced by the timing and shape of the ventilator waveform.   Since alveolar pressure cannot be measured directly it is estimated from a mechanical model of the lungs.   One such approach by Jain and Guha (1972) used rectangular pulses, with fixed inspiratory time and pressure from the pump, and the controller designed to adjust tidal volume by varying the expiratory time.

6.   ANTI-COAGULANT THERAPY

Anticoagulant drugs such as Warfarin are used to combat thrombosis and reduce blood clotting by interfering with the synthesis of prothrombin, or plasma protein whose activity controls the formation of fibrin through a chain of steps.   The goal is to reduce prothrombin without causing bleeding, by designing a suitable drug therapy profile.

An early approach by Sheiner (1969) planned the Warfarin dosage starting from the fourth day of therapy.   The initial dosage was calculated from the estimated body surface area of the patient, the second dose was half the initial dose, and no dose was given on the third day.   Theofanous and Barile (1973) extended the technique to cover several days of future drug applications, and presented formulas for both uniform and quasi-uniform (uniform except for a single initial loading) drug schedules. Four parameters characterise the patient according to two differential equations, and the parameters, which are constrained to lie within reasonable values, are determined by nonlinear least squares data fitting. The methods were evaluated   retrospectively using data from patients receiving conventional anticoagulant therapy, and were found to perform better than some of the cardiologists.

Powers et al (1980) represented the effect of Warfarin on prothrombin activity by a set of three state variables (G, Q and P) which were linked by three first-order state equations.   The three equations contained four parameters, but two parameters which do not strongly affect the prothrombin complex activity were assumed to be constant.   By computing two state variables on the basis of the constants and assuming initial conditions, only one state variable (P) and two constants were actually estimated from the data using extended Kalman filtering.   The filter gave faster identification than that using nonlinear least squares, but the microcomputer using BASIC was not suitable for calculation of optimal dosage strategies.

7.   COMPUTER-AIDED DRUG THERAPY

In addition to on-line control in biomedicine, computers have been used as consultants giving advice on drug therapy.   One example of this is the use of digoxin for many forms of heart failure.   Careful control is particularly important in this case because of the narrow margin between therapeutic and toxic dosages for digoxin.   Early work in this area by Jelliffe et al (1970) used a one-compartment model to describe digitalis pharmacokinetics in patients with normal and reduced renal function. Dose regimes to meet the required therapeutic goal (plasma concentration of the drug) were calculated from data on past history of dosages, renal function and body weight.

Sheiner et al (1972) introduced feedback into the above scheme using serum digoxin concentration as the modelled variable.   In this approach maximum likelihood estimation was used to determine the pharmacokinetic model parameters.   Although claiming improved performance over physicians, (Sheiner et al (1975)), some studies have shown that many cases of digitalis toxicity are not related to impaired renal function and this would not be avoided by use of such a kinetic model.   To encapsulate the many facets of expert clinical knowledge a digitalis management programme has been developed by Gorry et al (1978).   This programme uses a formal model of pharmacokinetics and a variety of qualitative clinical data to construct a patient-specific model for which the dosage regime is determined.   These recommendations are revised in a feedback loop based on the patient's clinical responses.   The programme has been evaluated retrospectively on clinical data in 19 patients.   In the 12% of cases which showed toxicity the patient received a higher dose than that recommended by the programme, while it also recognised the toxicity prior to its recognition by the clinician.

8.   EXPERT CONTROL

In addition to the current interest in expert systems for diagnostic purposes, there is also the possibility of using these ideas for on-line control of drug administration.   In such work, an attempt is made to include some form of human expertise or heuristic into the overall control system.   An early example of such an approach was that of Sheppard who used a 'decision table' in conjunction with a PI controller to keep the drug input consistent with a number of imposed clinical and physiological constraints.   In this way, the control system becomes effectively nonlinear in behaviour, but can cope with a wider range of patient conditions.

The introduction of expert system methodology into the management of the critically ill has been considered by Cramp and Carson (1986a) in the so-called Bedside Intelligent System (BIS).   In this case large amounts of laboratory data must be merged with other clinically relevant information to provide refined decision-making aids from which the clinician can determine suitable therapeutic dosage regimes.   This concept has been developed further for the particular case of control of fluid balance (Cramp and Carson, 1986b).

It is known that humans can control systems in a feedback manner very effectively even when they have only a rudimentary awareness of the model representing the object they are controlling.   It has been observed that in determining control actions the human being uses a method of reasoning which is apparently imprecise, but effective.   In line with this idea, the concept of fuzzy logic was introduced by Zadeh and has been studied extensively in industrial applications.   More recently, fuzzy logic control has been considered for muscle relaxant administration by Linkens and Mahfouf (1988).   Successful control was established via relatively simple fuzzy knowledge-based rules after considerable experimentation with a simulated model.   The technique used was to train a human to control a computerised model with suitably randomised parameters.   The logged results of these simulations were then analysed to extract the knowledge representing the human rule-base which has been established via the training sessions.   These rules were then coded into the fuzzy expert system and re-validated.   A similar procedure was

used in operating theatre based on logging data from an anaesthetist performing manual on-line control of drug infusion (Atracurium in this case)
Some work has been done in systems of self-organising fuzzy control,
whereby the rules are modified automatically according to the current
performance of the on-line system.   This has recently been extended to
provide fuzzy control of muscle relaxant anaesthesia without prior knowledge of rules (Linkens and Hasnain, 1989).   Fuzzy control has also been
applied to blood pressure management (Yamashita and Suzuki, 1988;   Ying
et al (1988), and to gaseous anaesthesia delivery (Vishnoi and Gingrich,
1987).

The use of expert systems methodology to provide an advisor for unconsciousness control in operating theatre has been studied using a graphical language GEM (Linkens et al, 1987).   This merges data from on-line
monitors and clinical observations to provide advice both on anaesthetic
state and recommended drug administration.   The knowledge base currently
comprises about 400 elements (cf rules), and uses Bayesian inference for
reasoning with uncertain (ie noisy) data.   It has undergone an iterative
development cycle involving a number of clinical evaluations (Linkens et
al, 1988).   The expert system is written in 'C', and a knowledge acquisition interface is being investigated using transputer parallelism
which segments the rule-base into 'packets', allowing easier editing of
the knowledge base.   Considerable attention has been given to the Human/
Computer Interaction aspects of the system.   The system has undergone a
number of small-scale evaluations, during which patient-variables and
anaesthetic dosages have been logged.

Another expert system has been applied in an advisory mode to the management of circulatory shock (Syed et al, 1987).   This rule-based system
monitors 14 patient parameters and manages heart rate, mean arterial
blood pressure, pulmonary artery wedge pressure, cardiac output, urine
flow rate and haematocrit.   Advice is provided on therapeutic interven-
tions by selection from 11 pharmacological agents, five fluids and two
devices.   Closed-loop management of cardiac shock using the expert-
system approach has not yet been reported;   however, investigation and
clinical testing of such systems are in progress.

Closed-loop management of mechanical ventilation is another area in which
the expert system may offer advantages.   Closed-loop control of ventilation has been carried out as part of the systems developed for controlling anaesthesia.   In these applications, management of blood oxygen
(PaO_2) and carbon dioxide ($PaCO_2$) partial pressures have been attempted
by monitoring blood oxygen saturation (SaO_2) using an oximeter and end-
tidal carbon dioxide ($ETCO_2$) using a capnograph.   Inspired oxygen concentration or respiratory rate and tidal volume are varied on the ventilator to hold SaO_2 and $ETCO_2$ at levels which give the correct level of
blood gases (PaO_2, $PaCO_2$).

Problems arise in such control loops because the relationship between
SaO_2 and PaO_2 and between $ETCO_2$ and $PaCO_2$ are nonlinear and depend on
the patient's physiological condition.   Skilled clinical decisions are
required to determine the appropriate reference levels of SaO_2 and $ETCO_2$
for the control-loop so that the desired PaO_2 and $PaCO_2$ levels are maintained.

9. CONCLUSIONS

The above applications involving automated drug administration have demonstrated that a wide range of control techniques are currently being explored in biomedicine. These range from simple on-off control to complex adaptive control schemes. Classical PID three-term controllers are commonly being used, but suffer from robustness problems when the patient dyamics are either unknown or time-varying. Biomedical systems often include pure time delays and this is encouraging the use of controllers based on the Smith predictor principle. Parameter estimation is an integral part of many adaptive systems, and this may become an important component in the control scheme where knowledge of drug dynamics is a desirable biproduct of regulation.

Measurement of a suitable variable which reflects closely the physiological state to be controlled is commonly a difficult matter. As a result of this, open-loop control based on a predetermined model is sometimes mandatory. The inadequacies of such an approach must, however, be circumvented using techniques such as feedforward and infrequent feedback of sampled information. The majority of applications use linear controller designs, since this reflects the current field of major knowledge in systems theory. Since nonlinearity is endemic in biological systems it is clearly desirable that attention should be given to improved design of controllers for such cases.

It is clear that the field of drug administration in biomedicine is fruitful for systems technologists and is stretching the available methods for the design of adequate feedback regulators. Each control strategy will have to be tailored to the particular application because of the complexity of living dynamic systems. Such tailoring will require inter-disciplinary knowledge of systems theory and physiological behaviour. The frontiers of knowledge in both areas should be advanced by such studies.

It is clear that in biomedical control applications the use of computers is now mandatory. Apart from the need for real-time computing in terms of control algorithm synchronisation, the requirements for operator interaction and patient monitoring are particularly crucial in this area. One example of this is in the management of intracranial pressure in patients with severe head injury in an intensive care unit (Mason, 1987). The inclusion of alarm signalling is even more critical than in process control, and the multiple requirements even for control of a relatively small subsystem show the need for well designed and portable multitasking software. This area will require considerable attention in the coming years.

It can be seen from the examples which have been discussed that in applying closed-loop computer control to patient care there are a number of factors which are common to all such applications and which can cause difficulties for the designer. A review of these difficulties can also help identify where effort needs to be directed:

(i) Patient complexity and variability: Human physiological systems are complex, highly interactive and nonlinear in their response to therapeutic infusions or outside influences. Modelling of such systems to provide a basis for designing the more complex multivariable controllers

is extremely difficult.   Expert systems based on clinical management
strategies and experience are likely to offer the best prospects for developing effective adaptable controllers.

(ii) Data collection: Effective control requires reliable and accurate
data collected at a rate which is consistent with the time constants of
the controlled system.   Online collection of data from implanted transducers is a requirement for the effective control of many patient systems.   Such transducers need to be biocompatible, robust, small and non-
disruptive.   Online data about parameters such as cadiac output, blood
gases and blood glucose levels are vital for the effective implementation
of closed-loop patient care.   While there are many exciting ideas under
investigation which employ the latest microelectronics and optical-fibre
technology to measure these parameters, there are still biocompatibility
and environmental problems which need to be overcome.   Much more attention needs to be focussed on this area.

(iii) Noise: Physiological noise arising from patient disorders or emo-
tional states, and non-physiological noise induced in transducers by
patient movement, nursing procedures or medical treatment pose problems
in implementing control.   Noise filtering and invalid data recognition
are important requirements of the control algorithm.

(iv) Safety and reliability: Stringent requirements are placed on the
design of medical equipment particularly with respect to electrical
safety.   These statutory requirements are stipulated in Standards Association documents.   In addition, every possible care must be taken to
protect the patient against failure of any component part of the control
system and alert the medical staff to system malfunction.   Audible and
visual alarms are needed.   Visual display units (VDU) provide good facilities for the display of patient data and alarm states.   The control
system's patient monitoring facilities provide the possibility of also
displaying on the VDU patient emergency states such as cardiac arrest.

(v)   User-interface facilities: The most ingenious and versatile controller will fail in the clinical environment if it does not have good
and appropriate user-interface facilities.   Medical and nursing staff
will reject any device which is difficult or tedious to use or which displays data in some manner which is unconventional in the medical sense
or difficult to interpret.   If the full potential of computer control
is to be realised in the medical environment, the design team must consist of control engineers and medical or nursing practitioners.   The
medical and nursing staff need to have an input to the design of the control strategy, the user interface and the hardware as well as participa-
ting in the clinical trials.

(vi) Modelling of physiological systems: Extensive models have been
developed for some physiological systems such as the cardiovascular system and the renal system.   However, one of the factors which is hamper-
ing the design and development of infusion controllers for vasoactive
drugs other than sodium nitroprusside is the lack of information on the
dynamics of the physiological response of the patient to changes in drug
infusion rate.   Similarly, the interactive effects caused by simultaneous multiple drug infusion need to be more thoroughly researched and
modelled.   This information is essential for designing suitable multi-
drug closed-loop control strategies and rigorously testing the proposed

hardware and software through simulation studies before the controller
is tested clinically.

It is evident that there is great potential for the application of
closed-loop computer control technology in medicine and particularly in
patient care.   There are problems to be overcome and some of these have
already been discussed.   There is one other which is possibly the greatest problem of all.   The fear of patient injury and the consequent liti-
gation is making the medical profession in some countries very reluctant
to utilise computer control even though they see its potential advantages.

This further emphasises the importance of designing safety-critical systems which consider not only the computer but also all other components
when designing for reliability and freedom from hazardous conditions.

REFERENCES

Ackerman E, Rosevear J W, and McGuckin W F, 1964, Phys in Med & Biol, 9,
203-213.

Albisser A M and Leibel B S, 1977, Clin Endocrin & Met, 6, 457.

Asbury A J, Henderson P D, Brown B H, Turner D J and Linkens D A, 1981,
Br J Anaesth, 53, 859-863.

Asbury A J, Linkens D A and Rimmer S J, 1982, Identification of interacting dynamic effects on muscle paralysis using Ethrane in the presence
of pancuronium, Br J Anaesth, 54, 790.

Beneken J E W, Blom J A, Jorritsma F F, Nandorff A, Bijnen A V and
Spierdijk J, 1979, Biomedizionische Technik, 24, 233.

Bickford R G, 1949, Neurophysiological applications of automatic anaesthetic regulator controlled by brain potential, Am J Physiol, 159, 562.

Bhansali P V and Rowley B A, 1984, J Clin Eng, 4751.

Brown B H, Asbury A J, Linkens D A, Perks R and Anthony M, 1980, Clin
Phys Physiol Meas, I, No 3, 203-210.

Cass N M, Lampard D G, Brown W A and Coles J R, 1976, Anaesth Intens
Care, 4, 16-22.

Chambille B, Guehard H, Londe M and Bargeton D, 1975, J Appl Physiol,
39, 837.

Chapman F W, Newell J C and Roy R J, 1985, Ann Biomed Eng, 13, 359-372.

Chilcoat R T, 1980, Trans Inst MC, Vol 2, No 1, 38-45.

Coles J B, Brown W A and Lampard D G, 1973, Med Biol Eng, 11, 262.

Coon R L, Zuperku E J and Kampine J P, 1978, Anaesthiology, 49, 201.

Cramp D G and Carson E R, 1986a, 3rd IMEKO Conf 'Measurement in Clinical
Medicine', Edin.

Cramp D G and Carson E R, 1986b, Biomed Meas Inf Cont 1.

Davies W L, Evans J M, Fraser A C L and Barclay M, 1982, Closed-loop control of anaesthesia, Symp on 'Control Systems Concepts and Approaches in
Clinical Medicine', Sussex, 5-7 April, 87-90.

East T D, Westenskow D R, Pace N L and Nelson L D, 1982, IEEE Trans Biomed Eng, BME 29, 736-740.

Frumin M J, 1956, Anaesthesiology, 18, 290.

Giard M H, Bertrand F O, Robert D and Pernier J, 1985, IEEE Trans Biomed Eng, BME 32, 658-667.

Gorry G A, Silverman H and Pauker S G, 1978, Am J Med, 64, 452.

Gregory V I, Katana P G and Chizeeks H G, 1987, IEEE Trans Biomed Eng, BME-34, 617-623.

Jacobs O L R, Bullingham R E S, Davies W L and Reasbeck M P, 1981, Feedback control of post-operative pain, IEE Conf Pub 194 'Control and its Applications', 52-56.

Jacobs O L R, Bullingham R E S, McQuay H J and Reasbeck M P, 1982, On-line estimation in the control of post-operative pain, 6th IFAC Symp on 'Identification and System Parameter Estimation', Washington, 7-11 June.

Jain V and Guha S K, 1972, IEEE Trans Biomed Eng, BME-19, 47.

Jazwinski A H, 1970, Stochastic Processes and Filtering Theory, Academic Press.

Jelliffe R W, Buell J, Kalaba R, Sridhar R and Rockwell R A, 1970, Math Biosci, 9, 179.

Kadish A H, 1964, The regulation and control of blood sugar level, Physiology and Cybernetic Simulation in 'Technicon International Symposium', New York, 82-85.

Kline N S, Shimano E, Stearns H, McWilliams L, Kohn M and Blair J H, 1968, Med Res Eng, Second Quarter, 14-19.

Koivo A J, 1980, IEEE Trans Biomed Eng, BME-27, 574-581.

Kondo K, Sano A, Kikuchi M and Sakurai Y, 1982, Adaptive and optimal blood glucose control system designed via state space approach, Inst MC Symp on 'Control Systems Concepts and Approaches in Clinical Medicine', Sussex, 5-7 April, 103-106.

Linkens D A, Asbury A J, Brown B H and Rimmer S J, 1981, Br J Anaesth, 53, 666P.

Linkens D A, Asbury A J, Rimmer S J and Menad M, 1982, Proc IEE, PtD, 129, 136.

Linkens D A, Asbury A J, Rimmer S J and Menad M, 1982, Self-tuning control of muscle relaxation during anaesthesia, IEE Conf on Applications of Adaptive and Multivariable Control', Hull, 19-21 July, 96.

Linkens D A, Denai M, Asbury A J, MacLeod A D and Gray W M, 1989, IFAC Conf 'Adaptive Systems in Control and Signal Processing', Glasgow, 381-386.

Linkens D A, Greenhow S G, Asbury A J and Rob H M, 1987, IEE Colloq 'Man-machine interface for IKBS', London.

Linkens D A, Greenhow S G, Asbury A J and Rob H M, 1988, BMIS Symp 'Expert Systems in Medicine, 5', London.

Linkens D A and Hasnain S B, 1989, Int Conf 'AI and Communicating Process Architectures' (eds M Reeve and S Ericsson Zenith), Imperial College, Wiley, 249-274.

Linkens D A and Mahfouf M, 1988, IFAC Symp 'Modelling and Control of

Biomedical Systems', Venice, 185-190.

Linkens D A and Mansour N E, 1989, IFAC Conf 'Adaptive Systems in Control and Signal Processing', Glasgow, 341-346.

McInnis B C and Deng L Z, 1985, Annals of Biomed Eng, 217-226.

Mansour N E and Linkens D A, 1989, IEE Proc PtD, 136, 1-11.

Mansour N E and Linkens D A, 1989, A Self-tuning Smith Predictor applied to the control of blood pressure, Comp Prog in Biomed.

Mapleson W W, Allott P R and Steward A, 1974, Br J Anaesth, 46, 805.

Mapleson W W, Chilcoat R T, Lunn J N, Blewett M C, Khatib M T and Willis B A, 1980, Br J Anaesth, 52, 234.

Mason J, 1987, Encyclopedia of Systems & Control, Pergamon Press.

Millard R K, Monk C R, Woodcock T E, Pereira E, Lewis G T R and Prys-Roberts C, 'Some applications of self-tuning control to blood pressure regulation', 1st IFAC-BME Symposium on 'Modelling and Control in Biomedical Systems', Venice, Pergamon, 89-100.

Moller D, Popovic D and Thiele G, 1983, in Irmfield Hartman (eds) 'Advances in Control Systems and Signal Processing, Vol 4'.

Packer J S, Mason D G, Cade J F and McKinley S M, 1988, IFAC Symp 'Modelling and Control in Biomedical Systems', Venice, 140-145.

Pickup J C and Keen H, 1980, Diabetologia, 18, 1-4.

Pickup J C, White M C, Keen H, Kohner E M, Parsons J A and Albert K G M M, 1979, Lancet II, 870-873.

Powers W F, Abbrecht P H and Covell D G, 1980, IEEE Trans Biomed Eng, BME-27, 520.

Reasbeck M P, 1982, Modelling and control of post-operative pain, DPhil Thesis, Oxford Univ.

Sano A, 1986, Biomed Meas Inf & Cont, 1, 16.

Serna V, Roy R and Kaufman H, 1983, JACC Conference.

Sheiner L B, 1969, Comp Biomed Res, 2, 507.

Sheiner L B, Rosenberg B and Melman K L, 1972, Comp Biomed Res, 5,441.

Sheiner L B, Halkin H, Peck C, Rosenberg B and Melman K L, 1975, Ann Intern Med, 82, 619.

Sheppard L C and Sayers B McA, 1977, Comp & Biomed Res, 10, 237-247.

Sheppard L C, Shotts J F, Robertson N F, Wallace F D and Kouchoukos N T, 1979, Proc First Ann Conf IEEE 'Eng in Medicine and Biology Soc', Denver, Colorado, Oct 6-7, 280-284.

Sirisena H R and Teng F C, 1986, Int J Syst Sci, 17.

Slate J B, Sheppard L C, Rideout V C and Blackstone E H, 1979, 5th IFAC Symp Ident & Sys Par Est, Darmstadt, FR Germany, Sept 24-28, 867-874.

Slate J B, 1980, Model-based design of a controller for infusing sodium nitroprusside during postsurgical hypertension, PhD Thesis, University of Wisconsin.

Smith J T and Schwede H O, 1972, Med biol Eng, 10, 207.

Suppan P, 1972, Br J Anaesth, 44, 1263.

Tatnall M L and Morris P, 1977, Simulation of halothane anaesthesia in neonates, in 'Biomedical Computing' (ed W J Perkins), Pitmans Med (UK).

Tatnall M L, Morris P and West P G, 1981, Br J Anaesth, 53, 1019.

Theofanous T G and Barile R G, 1973, J Pharm Sci, 62, 261.

Vishnoi R and Gingrich K J, 1987, Proc 26th Conf 'Decision & Control', Los Angeles.

Walker B K, Chia T L, Stern K S and Katona P G, 1982, IFAC Symp Ident & Syst Par Est, Washington, USA.

Wesseling K H, Settels J J, Walstra H G, Van Esch H J and Donders J J H, 1982, Proc 'App Phy to Med Biol' (eds G Alberi, Z Bajzer).

Westenkow D R, Bowman R J, Ohlson K B and Raemer D B, 1980, Med Instrum, 14, 311.

White W P, Pearce D J and Norman M, 1979, BMJ 2, 166-167.

Yamashita Y and Suzuki M, 1988, J Chem Eng (Japan), 21, 541-3.

Ying H, Sheppard L and Tucker D, 1988, Med Prog Through Tech, 13, 203-215.

Chapter 14

Robot control

A. S. Morris

1. INTRODUCTION

Robots offer the opportunity of automating manufacturing processes in a flexible way which has far reaching consequences on the economics of automation. They have made such a great impact because the effort required to reprogramme them, for instance to change the pattern of welds in a spot-welding application, requires only an adjustment to the controlling software. This is relatively simple compared with the mechanical redesign necessary when the functioning of hard automation machinery needs changing. The industrial robot is therefore of great importance in the modern manufacturing scene, and this is reflected in the fact that the application of robots to industrial tasks is growing at the rate of about 50% per year.

The Western definition of an industrial robot is that it is a multi-linked mechanical system in which the motion of the links is governed by a reprogrammable controller. This can be modelled by the system shown in figure 1, where the successive links form a chain with a joint between each pair of links. In figure 1, q_0 represents the base of the robot, which is usually bolted down at some known reference point on the factory floor, and q_n represents the tip of the end-effector. Each joint is known as a degree of freedom (DOF) and therefore the commonest form of industrial robot with six degrees-of-freedom has six links and six joints. The robot is normally constructed in two parts, a 3-DOF arm plus a 3-DOF end-effector (the tool on the end of the arm). This 6-DOF configuration is the minimum one which will allow the end-effector to be placed anywhere within the working envelope of the robot and at any angle with respect to the workpiece.

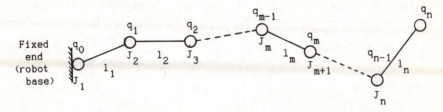

FIGURE 1 - TYPICAL CHAIN OF LINKS

2. REQUIREMENTS AND CONSTRAINTS FOR ROBOT MANIPULATOR CONTROL

The fundamental requirement in all robot applications is to be able to place the end-effector accurately in some defined position and orientat-

ion. In many tasks, there is also the additional requirement that the
end-effector should adhere closely to some specified trajectory through
its workspace as it moves between two such steady-state positions. The
parameters affecting a robot's ability to meet these requirements include its geometry, what trajectory is required, what drive system is
used, what computer power is available and what sensors are available.

Some of the problems involved in programming and controlling robots
start to become apparent if figure 1 is studied more closely. The system
of links comprising the robot is fixed at one end and free at the other.
This is a cantilever-like structure which is naturally oscillatory.
Further complication is added by the fact that the joint motions are not
in general all in the same plane. Therefore, the geometrical description of the robot makes the task of programming it to meet static control
purposes difficult. A more serious consequence, however, is that, as
the robot moves and its geometrical configuration changes, so the inertia
forces involved in its motion change. Dynamic control of the robot is
therefore made very difficult because of this time-varying nature of
some of the controlled parameters.

3.   MANIPULATOR GEOMETRY

Although in general a robot joint could be a screw one having both translational and rotational motion, all joints used in current production
robots give either pure translations or pure rotations only.

The positions of the joints (normally three) of the robot arm determine
the position of the end effector in the robot workspace. The motions of
the end-effector itself merely act about a fixed point in space and
affect the end-effector orientation.

The geometry of the robot is defined according to the type of joints
used in the arm. With the alternatives of translational or rotational
motion at each of three joints in the arm, there would appear to be four
possible geometrical arm configurations, as shown in figure 2. A fifth
configuration, SCARA, has recently appeared which has the same type of
joints as a polar-coordinate robot but arranged differently. Each type
of geometry is given a name:

Cartesian - 3T
Cylindrical - 2T, 1R
Polar - 1T, 2R
Revolute - 3R (also known as anthropomorphic or arm-and-elbow)
Scara - 1T, 2R (translational joint is normally pneumatic, driving
                between two fixed positions)

Each geometry gives the robot a particular working envelope. Flexibility
of motion and size of the working envelope increases as the number of
rotational joints in a robot increases, but so too does the programming
difficulty.

4.   DRIVE SYSTEMS

The five alternative drive systems available are dc motors, ac motors,
stepper motors, hydraulic actuators and pneumatic actuators. The relative merits of these can be summarised as follows:

DC motors   high accuracy, low maintenance, simple but low power
                          to weight ratio.

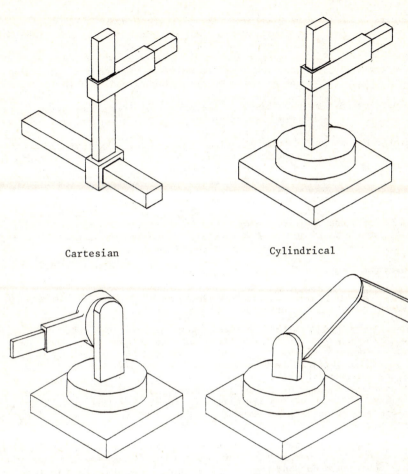

Cartesian   Cylindrical

Polar   Revolute

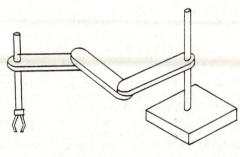

Scara

FIGURE 2 GEOMETRICAL ARRANGEMENTS

AC motors	high accuracy, very low maintenance, simple, now competitive on cost with dc motors but still have low power to weight ratio.
Stepper motors	very high accuracy, cheaper than dc/ac motors but very low power to weight ratio and can loose step for anything but very light loads due to load inertia.
Hydraulic actuators	high power to weight ratio, fair accuracy but messy with high maintenance cost and performance degradation over time.
Pneumatic actuators	cheap, reliable, medium power to weight ratio but dynamic performance problems due to compressibility of air and normally necessary to operate them between fixed stops (hence noisy).

To meet positioning accuracy requirements, electric motors are the clear favourite for the robot drive system, as the problems of friction and fluid leaks cause performance problems in pneumatic and hydraulic systems. However, the low power to weight ratio of electric drives precludes their use wherever the required payload of the robot is not small.

5. PROGRAMMING

Three methods of programming robots are currently in use, 'lead-through', 'drive-through' and 'direct'.

In the 'lead-through' method, a human operator moves the robot end-effector through the required sequence of motions manually. The robot control computer monitors the output of position sensors on the manipulator joints whilst this is happening and builds up a time-history of positions. Subsequently, the computer recalls this stored sequence of positions from memory and the manipulator follows the taught sequence of operations. Because of the limitations of human strength in moving the robot arm about, this programming technique is clearly limited to small light robots.

The 'drive-through' method is similar to the above but involves the robot being driven through the required sequence of motions by pressing buttons on a teaching pendant, with the control computer recording a history of joint positions as before. Teaching curved trajectories and moving to exact positions by this method is very difficult.

In 'direct' programming, target points and trajectories are defined to the robot controller as a series of (x,y,z) coordinate points. Provision must be made in the controlling software to convert these spatial (x,y,z) coordinates into robot joint positions, a procedure known mathematically as the inverse kinematic transformation. The degree of difficulty in doing this is a function of the number of rotational joints in the robot. For a cartesian coordinate robot, the effort is trivial but for a revolute coordinate robot it is very complicated. Significant errors always exist in calculating the position of the end-effector in spatial (x,y,z) coordinates from joint position measurements. Such errors arise because of deviations in the link-lengths etc., of robots away from nominal values at the time of their manufacture, because of joint position transducer errors and because of link flexure, backlash etc.

The minimum provision of sensors necessary for any of these three methods of programming and control are sensors to measure the positions of the robot joints. However, the quality of control is greatly enhanced if additional sensors are incorporated which provide information about the absolute position of the end-effector in space and/or its position with respect to other objects in the workspace.

6. SENSORS

The range of sensors found in industrial robots can be conveniently classified as internal and external sensors.

Internal sensors comprise those transducers which are concerned with monitoring joint motion and providing feedback information to the robot controller. This allows the controller to operate as a servomechanism and ensures that the required joint positions as defined by the inverse kinematic transformation are achieved. Transducers provide either position or velocity information. The usual devices used are:

Translational position - potentiometer, LVDT (linear variable differential transformer)
Rotational motion - potentiometer, encoder (especially optical form), synchro, resolver
Rotational velocity - dc tachogenerator, ac tachogenerator (less common)

External sensors are those sensors which provide information about the relationship between a robot and its environment and enable small positioning errors in the end-effector to be corrected. The range of sensors in this category includes:

range
proximity
touch
force
vision

As far as industrial applications are concerned, vision is by far the most important of these, with force sensing also being important to a lesser extent. Visions systems provide information about the identification of workpieces and their position and orientation with respect to the robot, and this information is valuable in most handling, processing and assembly tasks. Force sensing is generally used to provide information about the reaction forces between the end-effector and workpiece in assembly operations. The magnitude and direction of the reaction forces provides information about the relative orientation between the robot and workpiece and allows the appropriate correction to be carried out.

Where such external sensors are included within a robot system, provision must be made within the robot controller to handle the extra information provided.

7. KINEMATIC MODELLING

A pre-requisite for robot programming and control is knowledge of the kinematic model describing its geometry. The kinematic model describes the relationship between the robot joint positions and the spatial position of its end-effector in (x,y,z) coordinates and is associated with two mathematical transformations, the forward kinematic transformation and the inverse kinematic transformation.

The forward kinematic transformation involves the calculation of the end-effector (x,y,z) coordinates from a given set of joint positions $(q_1, q_2 \dots q_n)$. The <u>inverse kinematic transformation</u> involves the calculation of the set of joint positions $(q_1, q_2, \dots q_n)$ which will cause the end effector to be at some specified (x,y,z) position in space. This is the transformation which is of principle importance for both programming and control purposes.

If a manipulator involves more than two rotational joints with non-parallel axes, writing down the kinematic relationships 'longhand' in terms of the sine and cosine relationships between successive joints is not viable. Instead, it is necessary to define the relationships in terms of a strict framework known as homogeneous transformations. This can be demonstrated by considering the two-link manipulator shown in figure 3(a). In an xyz coordinate frame with origin at A, the position of the end of the arm C with respect to A is defined by the sum of two vectors r_1 and r_2 (figure 3(b)).

$$r_1 = (l_1\cos\theta_1), \quad (l_1\sin\theta_1), \quad (0)$$
$$r_2 = (l_2\cos(\theta_1+\theta_2)), \quad (l_2\sin(\theta_1+\theta_2), \quad (0)$$
$$r_1 + r_2 = (l_1\cos\theta_1 + l_2\cos(\theta_1+\theta_2)), \quad (l_1\sin\theta_1 + l_2\sin(\theta_1+\theta_2)), (0)$$

This is a simple case with only two joints. Also the axes are parallel. For more than two joints, with non-parallel axes, we need an alternative form of relationship.

What we need to do is to be able to express the position of any link in a coordinate frame whose origin is at the activating joint for the link. Such 'link' coordinate frames can be used providing that a transformation matrix can be synthesised which relates to the link coordinate frame to the reference coordinate frame, i.e. in the above example, express C in a coordinate frame with origin at B and relate this back to the reference coordinate frame at A by a transformation matrix.

Consider a coordinate frame uvw with origin at B (figure 3(c)). Let the coordinates of point C in this coordinate frame be (u_1, v_1, w_1).

Then:
$$\begin{aligned} x_1 \\ y_1 \\ z_1 \end{aligned} = R \begin{vmatrix} u_1 \\ v_1 \\ w_1 \end{vmatrix} \tag{1}$$

$$u_1 = l_2\cos\theta_2 \; ; \; v_1 = l_2\sin\theta_2 \; ; \; w_1 = 0$$

We require:
$$\begin{vmatrix} x_1 \\ y_1 \\ z_1 \end{vmatrix} = \begin{vmatrix} r_{11} & r_{12} & r_{13} \\ r_{21} & r_{22} & r_{23} \\ r_{31} & r_{32} & r_{33} \end{vmatrix} \cdot \begin{vmatrix} l_2\cos\theta_2 \\ l_2\sin\theta_2 \\ 0 \end{vmatrix} \tag{2}$$

such that
$$x_1 = l_1\cos\theta_1 + l_2\cos(\theta_1+\theta_2) \tag{3}$$

$$y_1 = l_1\sin\theta_1 + l_2\sin(\theta_1+\theta_2) \tag{4}$$

$$z_1 = 0 \tag{5}$$

From (2)
$$x_1 = l_2 (r_{11}\cos\theta_2 + r_{12}\sin\theta_2) \tag{6}$$

Comparing (3) and (6), we see that whatever we choose for r_{11} or r_{12}, we cannot obtain the $(l_1\cos\theta_1)$ term without getting unwanted terms in the product $(l_1 l_2)$. Thus R cannot be expressed as a 3*3 matrix. This is because R has to describe the translation from A to B as well as the

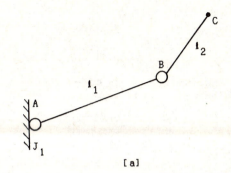

[a]

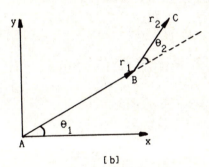

[b]

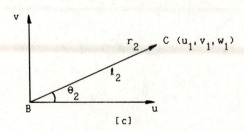

[c]

FIGURE 3

rotation of the uvw frame about the z axis with respect to the xyz frame. In fact, R has to be a 4*4 matrix to do this and equation (1) is modified to:

$$\begin{vmatrix} x_1 \\ y_1 \\ z_1 \\ 1 \end{vmatrix} = R \cdot \begin{vmatrix} u_1 \\ v_1 \\ w_1 \\ 1 \end{vmatrix} \qquad (7)$$

This is known as the homogeneous coordinate representation and matrix R is known as a homogeneous transformation.

If R is chosen as
$$\begin{vmatrix} \cos\theta_1 & \sin\theta_1 & 0 & 1_1\cos\theta_1 \\ \sin\theta_1 & \cos\theta_1 & 0 & 1_1\sin\theta_1 \\ 0 & 0 & 1 & 0 \\ 0 & 0 & 0 & 1 \end{vmatrix}$$

it can be shown that equation (7) is satisfied.

This method of transformation between coordinate frames can be extended to any number and configuration of links. Standard homogeneous transformations exist which describe single translations or rotations between successive coordinate frames. These act as building blocks which enable any transformation matrix to be synthesised by multiplying together the appropriate standard transformations. Six such standard homogeneous transformations exist:

8. STANDARD HOMOGENEOUS TRANSFORMATIONS

Translation by a distance a along the x axis:
$$\text{Trans }(a,0,0) = \begin{vmatrix} 1 & 0 & 0 & a \\ 0 & 1 & 0 & 0 \\ 0 & 0 & 1 & 0 \\ 0 & 0 & 0 & 1 \end{vmatrix} \qquad (8)$$

Translation by a distance b along the y axis:
$$\text{Trans }(0,b,0) = \begin{vmatrix} 1 & 0 & 0 & 0 \\ 0 & 1 & 0 & b \\ 0 & 0 & 1 & 0 \\ 0 & 0 & 0 & 1 \end{vmatrix} \qquad (9)$$

Translation by a distance c along the z axis:
$$\text{Trans }(0,0,c) = \begin{vmatrix} 1 & 0 & 0 & 0 \\ 0 & 1 & 0 & 0 \\ 0 & 0 & 1 & c \\ 0 & 0 & 0 & 1 \end{vmatrix} \qquad (10)$$

Rotations by an angle θ:

About X axis:
$$\text{Rot }(x,\theta) = \begin{vmatrix} 1 & 0 & 0 & 0 \\ 0 & \cos\theta & -\sin\theta & 0 \\ 0 & \sin\theta & \cos\theta & 0 \\ 0 & 0 & 0 & 1 \end{vmatrix} \qquad (11)$$

About y axis:
$$\text{Rot }(y,\theta) = \begin{vmatrix} \cos\theta & 0 & \sin\theta & 0 \\ 0 & 1 & 0 & 0 \\ -\sin\theta & 0 & \cos\theta & 0 \\ 0 & 0 & 0 & 1 \end{vmatrix} \qquad (12)$$

About z axis: Rot (z,θ) = $\begin{vmatrix} \cos\theta & -\sin\theta & 0 & 0 \\ \sin\theta & \cos\theta & 0 & 0 \\ 0 & 0 & 1 & 0 \\ 0 & 0 & 0 & 1 \end{vmatrix}$ (13)

We can now see how these were used to formulate R for equation (7) earlier. Transforming between the two coordinate frames with origins at A and B involves a translation by a distance l_1 along the x axis and a rotation by an angle θ_1 about the z axis.

Thus R = $Rot(z,\theta_1)$. $Trans(l_1,0,0)$
Using equations (8) and (13):
R = $\begin{vmatrix} \cos\theta_1 & -\sin\theta_1 & 0 & 0 \\ \sin\theta_1 & \cos\theta_1 & 0 & 0 \\ 0 & 0 & 1 & 0 \\ 0 & 0 & 0 & 1 \end{vmatrix}$. $\begin{vmatrix} 1 & 0 & 0 & l_1 \\ 0 & 1 & 0 & 0 \\ 0 & 0 & 1 & 0 \\ 0 & 0 & 0 & 1 \end{vmatrix}$ = $\begin{vmatrix} \cos\theta_1 & -\sin\theta_1 & 0 & l_1\cos\theta_1 \\ \sin\theta_1 & \cos\theta_1 & 0 & l_1\sin\theta_1 \\ 0 & 0 & 1 & 0 \\ 0 & 0 & 0 & 1 \end{vmatrix}$

In general n-link system, the coordinates of point P_m can be related to a coordinate frame with an origin at P_{m-1} by a transformation matrix A_m.

i.e. $P_m = A_m . P_{m-1}$ (14)
Similarly, points P_{m-1} and P_{m-2} are related by:

$P_{m-1} = A_{m-1} . P_{m-2}$ (15)

Combining (14) and (15): $P_m = A_{m-1} . A_m . P_{m-2}$ (16)
Continuing in a similar fashion: $P_m = A_1.A_2.A_3....A_{m-1}.A_m.P_0$ (17)
P is the origin of the base coordinate frame and can be expressed in homogeneous coordinates as: $P_0 = (0\ 0\ 0\ 1)^T$ (18)

It is conventional to express the matrix product which relates the end of the manipulator chain P_0 to the base coordinate frame as the quantity T i.e. for an n-link manipulator, $T_n = A_1 A_2 .... A_n$ (19)
For a 6-DOF manipulator, $T_6 = A_1 A_2 A_3 A_4 A_5 A_6$ (20)

To apply these relationships correctly, it is important that the axes of all coordinate frames involved are defined consistently. One proven set of rules for this is the Denavit-Hartenburg system.

9. DENAVIT-HARTENBURG MANIPULATOR REPRESENTATION[1]

This establishes an orthonomal cartesian coordinate system for each link in the manipulator, with the origin of each coordinate frame at the joint axes. Coordinate frame i (i=1...n for an n-link system) has its origin at joint (i+1) and so the frame moves as the link moves.

For each link i, the corresponding coordinate frame i has unit vectors along its principal axes given by (x_i, y_i, z_i). An additional coordinate frame O is defined with its origin at the robot base (the fixed end of link 1) with unit vectors given by (x_0, y_0, z_0) along its principal axes. Thus for an n-link manipulator, there are n+1 coordinate frames.

Every coordinate frame is defined according to three rules:

(i)   The z_{i-1} axis lies along the axis of motion of the ith joint.
(ii) The x_i axis is normal to the z_{i-1} axis and pointing away from it.
(iii) The y_i axis completes a right-handed coordinate system.

Any rotational or translational joint can be described by a set of four geometric parameters associated with its corresponding link. These four parameters are:

(a) θ_i the joint angle when turning about the z_{i-1} axis and moving from the x_{i-1} axis to the x_i axis (using the right-hand rule).

(b) d_i the distance along the z_{i-1} axis from the (i-1) coordinate frame origin to the intersection with the x_i axis.

(c) a_i the offset distance along the x_i axis from the ith coordinate frame origin to the intersection with the z_{i-1} axis. (i.e. the shortest distance between the z_{i-1} and z_i axes).

(d) α_i the offset angle when turning about the x_i axis and moving from the z_{i-1} axis to the z_i axis (using the right-hand rule).

For a rotary joint, d_i, a_i and α_i remain constant, with θ_i being the joint variable.

For a translational joint, θ_i, a_i and α_i remain constant, and d_i is the joint variable.

10. KINEMATIC EQUATIONS OF A REAL ROBOT

Before going on to derive the kinematic relationships for a real robot, we need to define some rules about the manner in which coordinate frames are defined and also define some sign conventions.

A homogeneous transformation matrix relating link (i-1) to link i can be synthesised as the product of the transformation matrices representing the following successive motions:

Rotate about z_{i-1} through an angle θ_i

Translate a distance d_i along z_{i-1}

Rotate about x_i through an angle α_i

i.e. $A = Rot(z,\theta) . Trans(0,0,d) . Trans(a,0,0) . Rot(x,\alpha)$

$$
= \begin{vmatrix} \cos\theta & -\sin\theta & 0 & 0 \\ \sin\theta & \cos\theta & 0 & 0 \\ 0 & 0 & 1 & 0 \\ 0 & 0 & 0 & 1 \end{vmatrix} \begin{vmatrix} 1 & 0 & 0 & 0 \\ 0 & 1 & 0 & 0 \\ 0 & 0 & 0 & d \\ 0 & 0 & 0 & 1 \end{vmatrix} \cdot \begin{vmatrix} 1 & 0 & 0 & a \\ 0 & 1 & 0 & 0 \\ 0 & 0 & 1 & 0 \\ 0 & 0 & 0 & 1 \end{vmatrix} \cdot \begin{vmatrix} 1 & 0 & 0 & 0 \\ 0 & \cos\alpha & -\sin\alpha & 0 \\ 0 & \sin\alpha & \cos\alpha & 0 \\ 0 & 0 & 0 & 1 \end{vmatrix}
$$

$$
= \begin{vmatrix} \cos\theta & -\sin\theta.\cos\alpha & \sin\theta.\sin\alpha & a\cos\theta \\ \sin\theta & \cos\theta.\cos\alpha & -\cos\theta.\sin\alpha & a\sin\theta \\ 0 & \sin\alpha & \cos\alpha & d \\ 0 & 0 & 0 & 1 \end{vmatrix} \tag{21}
$$

Consider now the six-axis Puma robot arm. This is a typical mixed-application revolute-coordinate industrial robot manufactured by Unimation Ltd. The six transformation matrices required to solve equation (20) can be found by substituting the parameter values for the PUMA robot in equation (21):

$$
A_1 = \begin{vmatrix} \cos\theta_1 & 0 & -\sin\theta_1 & 0 \\ \sin\theta_1 & 0 & \cos\theta_1 & 0 \\ 0 & -1 & 0 & 0 \\ 0 & 0 & 0 & 1 \end{vmatrix} \qquad A_4 = \begin{vmatrix} \cos\theta_4 & 0 & -\sin\theta_4 & 0 \\ \sin\theta_4 & 0 & \cos\theta_4 & 0 \\ 0 & -1 & 0 & 433 \\ 0 & 0 & 0 & 1 \end{vmatrix}
$$

$$
A_2 = \begin{vmatrix} \cos\theta_2 & -\sin\theta_2 & 0 & 432\cos\theta_2 \\ \sin\theta_2 & \cos\theta_2 & 0 & 432\sin\theta_2 \\ 0 & 0 & 1 & 149 \\ 0 & 0 & 0 & 1 \end{vmatrix} \qquad A_5 = \begin{vmatrix} \cos\theta_5 & 0 & \sin\theta_5 & 0 \\ \sin\theta_5 & 0 & -\cos\theta_5 & 0 \\ 0 & 1 & 0 & 0 \\ 0 & 0 & 0 & 1 \end{vmatrix}
$$

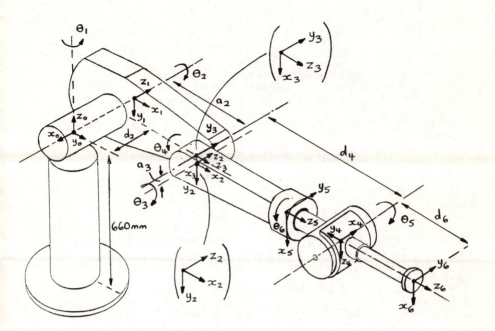

Parameters for PUMA 560 robot					
Joint	θ	α	a	d	joint range
1	+90	-90	0	0	-160 to +160
2	0	0	432mm	149mm	-225 to +45
3	+90	+90	-20.5mm	0	-45 to +225
4	0	-90	0	433mm	-110 to +170
5	0	+90	0	0	-100 to +100
6	0	0	0	56mm	-266 to +266

FIGURE 4 PUMA ROBOT

$$A_3 = \begin{vmatrix} \cos\theta_3 & 0 & \sin\theta_3 & -20.5\cos\theta_3 \\ \sin\theta_3 & 0 & -\cos\theta_3 & -20.5\sin\theta_3 \\ 0 & 1 & 0 & 0 \\ 0 & 0 & 0 & 1 \end{vmatrix} \quad A_6 = \begin{vmatrix} \cos\theta_6 & -\sin\theta_6 & 0 & 0 \\ \sin\theta_6 & \cos\theta_6 & 0 & 0 \\ 0 & 0 & 1 & 56 \\ 0 & 0 & 0 & 1 \end{vmatrix}$$

We have now established a mechanism for calculating the forward kinematic transformation which gives the position of the end effector from known joint angles. i.e. we have found a matrix T such that:

$$v = T \cdot q \tag{22}$$

where v is the vector of (xyz) coordinates and q is the vector of joint angles.

However, this gives no information yet about the end-effector orientation which is needed on many occasions. It is inappropriate to discuss this here and the interested reader is referred to a suitable text e.g. Fu (1987)[2].

11. INVERSE KINEMATIC RELATIONSHIPS

The inverse kinematic problem is to calculate the set of joint positions which will give the correct end effector position specified in (xyz) coordinates. i.e. we need to find the inverse of T such that:

$$q = T^{-1} \cdot v \tag{23}$$

Extraction of the joint positions from this requires extremely lengthy mathematical manipulation which cannot be reproduced here, although it is presented in full elsewhere[3]. Comment should be made however that matrix relationship (23) yields 12 equations in the 6 unknowns $\theta_1 \ldots \theta_6$. This indicates that there are multiple solutions, which can be expected since T involves sine and cosine relationships which yield alternative positive and negative values for θ. There is no 'general case' inverse kinematic solution, therefore, and a separate solution has to be worked out for every different manipulator configuration.

12. DYNAMIC MODELS

The kinematic modelling covered so far describes the steady-state position of the robot end effector. Whilst moving between positions, however, velocity and acceleration related dynamic forces are in action and these cause excursions of the end-effector away from its target position/trajectory. In order to control such effects, a dynamic model of the system is required. The two main dynamic model formulations are the Newton-Euler and the Lagrange-Euler equations. Previous reviews of these and various other formulations which are available[4,5] have shown that only the Lagrangian algorithm is readily suited to controller synthesis and analysis.

The Newton-Euler approach is based on a set of forward and backward recursive equations which are applied to the links sequentially. Newton's translational motion equation together with its analogue, Euler's rotational motion equation, describe how forces, inertias and accelerations relate.

The forward recursion propagates kinematics information and calculates the velocities and accelerations at each joint. The backwards recursion then uses this information to calculate the forces and torques at each joint. The computational burden involved in doing this is very much less than that involved in the Lagrangian formulation. However, the procedure is messy and the equations are not in a form which is readily useable for the synthesis of advanced control algorithms. The full derivation of the Newton-Euler formulation can be found elsewhere[6].

The Lagrangian L is defined as the difference between the kinetic energy and potential energy of the robot system, and is expressed as a function involving all masses and link velocities in the system. This produces a set of second-order, coupled, non-linear differential equations.

The torque/force at each joint is calculated by differentiating the Lagrangian with respect to the corresponding joint positions and velocities. The equation so derived for the system torques/forces involves coupling inertia terms which describe forces on joints which arise due to the motion of succeeding and preceding joints along the manipulator chain. These are known as centripetal and coriolis forces. The force/torque at joint i is given by:

$$F_i = \sum_{j=1}^{n} P_{ij} \ddot{q}_j + \sum_{j=1}^{n} \sum_{k=1}^{n} P_{ijk} \dot{q}_j \dot{q}_k + P_i \tag{24}$$

where   for $i = 1, 2, \ldots n$

P_{ii} : effective inertia at joint i

P_{ij} : coupling inertia between joints i and j

$$= \sum_{\ell=max(i,j)}^{n} Tr \left[\frac{\partial H_\ell}{\partial q_j} \, J_\ell \, \left| \frac{\partial H_\ell}{\partial q_i} \right|^T \right] \tag{25}$$

P_{ijj}: centripetal forces at joint i due to velocity at joint j

P_{ijk}: coriolis forces at joint i due to velocities at joints j and k

$$= \sum_{\ell=max(i,j,k)}^{n} Tr \left[\frac{\partial^2 H_\ell}{\partial q_j \partial q_k} \, J_\ell \, \left| \frac{\partial H_\ell}{\partial q_i} \right|^T \right] \tag{26}$$

P_i : gravity loading vector

$$= \sum_{\ell=1}^{n} -m_\ell g^T \left| \frac{\partial H_\ell}{\partial q_i} \right| {}^\ell r_\ell \tag{27}$$

with $\dfrac{\partial H_\ell}{\partial q_i} = H^\ell \Delta_i \; ; \; m_\ell$ = mass of link ℓ;   (28)

${}^\ell r_\ell$ = centre of mass of link ℓ relative to its own coordinate frame

where H_ℓ is a 4 X 4 transformation matrix

and Δ_i^ℓ is the differential translation and rotation transformation matrix of joint ℓ with respect to the ith joint coordinate given by:

$$
{}^{\ell}\Delta_{iz} = \begin{vmatrix} 0 & -{}^{\ell}\delta_{iz} & {}^{\ell}\delta_{iy} & {}^{\ell}d_{ix} \\ {}^{\ell}\delta_{iz} & 0 & -{}^{\ell}\delta_{ix} & {}^{\ell}d_{iy} \\ -{}^{\ell}\delta_{iy} & {}^{\ell}\delta_{ix} & 0 & {}^{\ell}d_{iz} \\ 0 & 0 & 0 & 0 \end{vmatrix} \tag{29}
$$

and J_i is a pseudo inertia matrix, where the elements composing the matrix are the moments of inertia, cross product of inertia and first moments of each link i.e. :

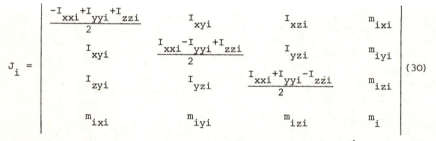

$$
J_i = \begin{vmatrix} \dfrac{-I_{xxi}+I_{yyi}+I_{zzi}}{2} & I_{xyi} & I_{xzi} & m_{ixi} \\ I_{xyi} & \dfrac{I_{xxi}-I_{yyi}+I_{zzi}}{2} & I_{yzi} & m_{iyi} \\ I_{zyi} & I_{yzi} & \dfrac{I_{xxi}+I_{yyi}-I_{zzi}}{2} & m_{izi} \\ m_{ixi} & m_{iyi} & m_{izi} & m_i \end{vmatrix} \tag{30}
$$

The establishment of this equation can be found elsewhere[4]. The derivation is simple and systematic but involves manipulation of 4*4 matrices which requires a very large computational effort. Solution of (24) takes approximately 200ms on a SUN workstation.

13.   ROBOT DYNAMIC CONTROL

The pre-requisite for robot control is for the required trajectory of the end-effector to be specified. Trajectory planning usually involves one of the following two possible tasks:

  (i) Design a path where the end-effector passes through a specified
       set of points in an unobstructed robot workspace.
  (ii) As (i) but with obstacles to be avoided in the workspace.

Any trajectory can be expressed as a succession of points and hence (i) above merely required solution of the inverse kinematic relationship (equation 23) for all defined points in the path or trajectory. Where obstacles are present in the workspace (case (ii) above), graphical search methods must be applied to establish the optimum path, expressed as a succession of (xyz) coordinates. Following this, equation (23) is applied again.

The control problem is then to move the end-effector (point q_n) along the specified trajectory. To synthesise a controller, knowledge of both the inverse and forward dynamic models is required. The inverse dynamic model (equation 24) calculates the forces and torques required to achieve some trajectory specified in terms of a set of q and $\dot{q}$ at successive intervals of time. The forward dynamic model calculates the values of q, $\dot{q}$ and $\ddot{q}$ at each joint resulting from the set of applied joint actuator forces and torques. A Lagrangian representation of the forward dynamics can be formulated by rearranging equation (24) in the following form:

$$\sum_{j=1}^{n} P_{ij}\ddot{q}_j = F_i - \sum_{j=1}^{n}\sum_{k=1}^{n} P_{ijk}\dot{q}_j\dot{q}_k - P_i \qquad (31)$$

$$\text{for } i = 1, 2, \ldots n$$

This represents a linear set of equations which can be solved by a Gaussian Elimination (GE) algorithm[7].

The first problem that arises in designing a controller for the robot is that parameters within equations (24) and (31) are dependent upon the magnitude of the load on the end of the robot end-effector. The load may or may not be readily definable, according to the task that the robot is performing. However, even more serious control difficulties arise from the fact that the gravity, centripetal, coriolis and inertia terms in equations (24) and (31) vary with the robot configuration and are therefore time-varying. This means that both the inverse and forward dynamic models must be computed at each control interval.

For accurate high speed control of typical manipulators, a sampling rate of at least 60Hz is required, i.e. the sampling interval must be less than or equal to 16ms. Indeed one author[8] has suggested that the sampling frequency should be at least 300Hz to allow for typical robot resonant frequencies around 15Hz. Clearly, the complexity of the dynamic model is such that it cannot be computed on-line with typical process control computers within such time constraints even in its simpler Newton-Euler form.

Robot manufacturers have met this problem by ignoring the cross-coupling terms and controlling each robot joint as separate single variable systems. In implementations of this strategy, a supervisory computer controller calculates target trajectories for each joint and passes the information to individual microprocessor-driven, servo-controllers on each joint which typically apply a three-term control law. For low speed arm motions, the cross-coupling terms are small and this approximation of the system to several single-variable ones is justified. However, to make this approximation valid, it is usually necessary to artificially limit joint speeds. This is contrary to the usual industrial requirement for robots to move as fast as possible.

For high speed motion, the cross-coupling terms are very significant and neglecting them leads to gross errors. Therefore, if satisfactory control of robots moving at higher speeds is to be achieved, the robot must be treated as a non-linear, multivariable system and controlled accordingly. Various approaches to synthesising a more accurate controller for high speed robot motion are being researched, as described below.

One control technique which has received a great deal of attention is the Computed Torque Scheme[9]. This involves calculating the joint forces and torques required to move the end-effector along some defined position, velocity and acceleration profile. This has the inherent advantage of incorporating feedforward control which improves dynamic performance. However, in general with all feedforward controllers, the control algorithm is very sensitive to unmodelled and inaccurately known parameters, e.g. friction forces, link lengths, link masses etc. Using conventional computing resources, the Newton-Euler formulation has to be used for the robot model involved in these calculations, as the Lagrange-Euler equations are too complex. However, recent research into transputer applications[10] has shown that transputers are capable of

computing the Lagrangian formulation of the dynamic model at the
required sampling rate and at a realistic cost in terms of computer
hardware.

Most programs of current research into robot control involve the development of various forms of adaptive controller. Koivo[11], for instance,
describes a Linear Adaptive Regulator which fits a linear autoregressive
model to robot input-output data. The controller is then synthesised
from the identified parameters. The main problem with this particular
scheme is that it assumes a linear model and hence the model coefficients
can only vary slowly compared with the controlled variables (q, $\dot{q}$ and $\ddot{q}$).
To satisfy this constraint, sampling rates much higher than those real-
istically achievable are demanded. A more promising approach is the
adaptive computed torque scheme[12]. This incorporates an adaptive
mechanism into the computed torque scheme and also has a proportional-
derivative feedback loop to take account of model inaccuracies and to
provide a disturbance rejection capability. Model Reference Adaptive
Controllers have also been investigated, by Han[13] using a nonlinear
reference model, and by Lim[14] using a linear reference model.

Other approaches to the problem are also being made. One such alternative method is variable structure control[15]. This is very tolerant of
dynamic model inaccuracies but may lead to control signals which change
sign rapidly, causing 'chattering' in the joint drive systems. Much
interest is also being shown in applying AI techniques[16] to learn a
control strategy based on measured input commands and robot responses.

A further active area of research concerns the control problems of
flexible manipulators[17]. Current robots are designed to have rigid links
where the degree of flexure is negligible. However, to be rigid, links
have to be of large section and hence of large mass, which greatly
increases inertia forces and produces corresponding dynamic control
problems. This area is worthy of further investigation because, whilst
flexure introduces control problems of its own, the usual inherent
control problems of large inertia forces are avoided.

14. CONCLUSIONS

This review of the state-of-the-art in robot control has highlighted the
great difficulties in synthesising a satisfactory controller. It has
been established that the manipulator is a nonlinear, multivariable
system with time-varying and/or inaccurately-known parameters. However,
most of the controllers proposed to date make simplifying assumptions
which treat the manipulator as a linear and/or single-variable system.
Such assumptions allow reasonably satisfactory control at low speeds but
cause gross errors in dynamic control at high speeds.

Various avenues of current research around the world are aimed at
developing a proper nonlinear and multivariable controller which is
tolerant towards parameter inaccuracies and variations. Presently,
results are being reported for simulated controllers and simulated robot
models. The following, and probably much more difficult stage, will be
to implement such control algorithms in a real robot system.

REFERENCES

1. Denavit, J. and Hartenburg, R.S.,: 'A kinematic notation for lower-
    pair mechanisms based on matrices' J. Appl. Mech., 22, 1955, pp215-
    222.

2. Fu, K.S., Gonzales, R.C. and Lee, C.S.G.,: 'Robotics: control, sensing, vision and intelligence', McGraw-Hill, 1987.

3. N-Nagi, F. and Siegler, A.: 'Engineering foundations of robotics', Prentice-Hall, 1987.

4. Paul, R.P.,: 'Robot manipulators: Mathematics, Programming and Control', MIT Press, 1981.

5. Zomaya, A.Y.H. and Morris, A.S.,: 'On the complexity reduction of the coriolis and centripetal effects of a 6-DoF robot manipulator', Proc of Int Workshop on Robot Control, Oxford, U.K., 1988 (published as Robot Control: theory and applications by Peter Peregrinus, pp 71-81).

6. Craig, J.J.,: 'Introduction to Robotics: Mechanics and Control', Addison-Wesley, 1986.

7. Press, W.H. et al.,: 'Numerical recipes: The art of scientific computing', Cambridge University Press, 1986.

8. Nigam, R. and Lee, C.S.G.,: 'A multiprocessor-based controller for the control of mechanical manipulators', IEEE J of Rob and Autom, RA-1, 1985, pp 173-182.

9. Lee, C.S.G, et al, Proc 6th IFAC conf on estimation and parameter identification, June 1982, pp 1154-1159.

10. Zomaya, A.Y.H. and Morris, A.S.,: 'Distributed VLSI architectures for fast kinematics and dynamics computations', Proc of IMA Conf on Robotics, Loughborough, U.K., July 1989 (to be published by Oxford University Press).

11. Koivo, A.J.,: 'Force-position-velocity control with self-tuning for robotic manipulators', IEEE Conf on Robotics and Automation, San Francisco, 1986.

12. Slotine, J.E. and Li, W.,: 'Adaptive manipulator control: a case study', Trans IEEE, vol AC-33, No.11, 1988.

13. Han, J-Y, et al,: 'Nonlinear adaptive control of an n-link robot with unknown load', Robotics Research, Vol 6, No.3, 1987

14. Lim, K.Y. and Eslami, M.,: 'Robust adaptive controller designs for robot manipulator systems', IEEE J of Rob and Autom, Vol RA-3, No.1 1987.

15. Young, K.K.D.,: 'Controller design for a manipulator using theory of variable structure systems;, Trans IEEE, SMC-8, 1978, pp 101-109.

16. Miller, W.T.,: 'Sensor-based control of robotic manipulators using a general learning algorithm', IEEE J of Rob and Autom, Vol RA-3, No.2, 1987.

17. Sasiadek, J.Z, and Srinivasan, R.,: 'Model reference adaptive systems for flexible manipulators', Robot Control: theory and applications, Peter Peregrinus, 1988, pp 71-81.

Chapter 15

Active control of a modern fighter aircraft

D. McLean

1.   INTRODUCTION

1.1   Fighter aircraft characteristics

Fighter aircraft are intended for use in air combat. They are used to engage an enemy's aircraft and missiles. They must fly fast and high and are extremely manoeuvrable. Having such characteristics results in their possessing distinctive flying qualities which profoundly affect the degree of success which fighter pilots can achieve in aerial combat. The need for higher speeds and improved manoeuvrability results in the aircraft's weight being low; whenever the aircraft is subjected to manoeuvre commands, or encounters atmospheric turbulence, its wing and fuselage will bend. Furthermore, if the fighter is to avoid detection by the enemy's radar, a great deal of composite material must be used in its construction instead of metal, and the use of such composite material results in the vehicle's weight being much lighter. In addition, this need to avoid radar detection can result in such aircraft having to be flown, in certain missions, very low at high speed. Such a flying tactic can then result in the aircraft's experiencing considerable unwanted acceleration, which can result in its pilot being unable to carry out his task satisfactorily. To overcome such problems modern fighter aircraft employ Active Control Technology (ACT).

1.2   Active control technology

Active control technology is the use of a multi-variable, automatic, flight control system (AFCS) to improve the dynamic flight characteristics, the manoeuvrability and, often, the structural dynamic properties of an aircraft by simultaneously driving several control surfaces, and auxiliary force and moment generators, in a manner that much reduces those loads which the aircraft would have experienced as a result of its motion in the absence of an ACT system, or causes the aircraft to produce a degree of manoeuvrability much beyond the capability of a conventional aircraft. Its purpose is to provide AFCS with the additional means of increasing the performance and the operational flexibility of the basic aircraft. As a result of the missions they must perform, the current requirements for modern fighter aircraft are such that the resulting configurations are greatly altered from the familiar designs of earlier times. In meeting the new requirements, these new designs have featured:

thin lifting surfaces

long slender fuselages

low mass fraction structures

low load factors

and a higher level of stress has been permitted.

These features have resulted in aircraft which are of the required structural lightness, but which exhibit, as a consequence, considerable flexibility. Such aircraft can develop structural displacements and accelerations of large amplitude.

This paper deals with the use of ACT on modern fighter aircraft to provide the fighter with the following flight control modes, viz:

stability augmentation

gust load alleviation

bending mode control

and   ride control.

The type of modern fighter, which has such systems, is one which was designed initially with ACT applications in mind. Typically, it would have many more control surfaces than a conventional aircraft. (see Figure 1). Such aircraft are referred to sometimes as Control Configured Vehicles (CCV).

1.3   ACT functions

The purpose of the stability augmentation system (SAS) is to provide the basic aircraft with additional damping, usually for short-period motion or to suppress the Dutch roll motion. Bending mode control, sometimes referred to as structural mode control, or active lift distribution control, is a technique of redistributing the lift generated by the wing of an aircraft during a manoeuvre. By the symmetrical deflection of control surfaces, located at proper stations on the trailing edge of the wing, in response to commands which cause a change in load factor, it is possible to reduce the incremental stress by arranging for the centre of lift of the wing to shift inboard. This inward shift also reduces the bending moment at the wing's root, a major factor in the fatigue life of any wing.

The purpose of the ride control system (RCS) is to improve the comfort of the crew of the fighter aircraft by reducing the objectionable levels of acceleration which are caused by the aircraft's motion. Finally, a gust load alleviation (GLA) system is a means of controlling the contribution of the rigid-body and bending modes to the complete dynamic response of a fighter aircraft whenever it encounters a gust or is subjected to a manoeuvre command. Its principal purpose is to reduce the transient peak loads which arise from such turbulence.

2.   MATHEMATICAL MODEL OF A FIGHTER AIRCRAFT

2.1   Rigid body motion

Using the stability axis system in which there is no steady vertical velocity and the steady value of Θ is defined as the equilibrium flight path angle, γ_o, the equations of small perturbed motion in the longitudinal plane may be expressed as:-

$$\left.\begin{aligned}
\dot{u} &= X_u u + X_w w - g\cos\gamma_o \theta + X_{\delta_{TH}}\delta_{TH} \\
\dot{w} &= Z_u u + Z_w w + U_o q - g\sin\gamma_o \theta + Z_{\delta_E}\delta_E + Z_{\delta_{TH}}\delta_{TH} \\
\dot{q} &= M_u u + M_w w + M_{\dot{w}}\dot{w} + M_q q + M_{\delta_E}\delta_E + M_{\delta_{TH}}\delta_{TH} \\
\dot{\theta} &= q
\end{aligned}\right\} \tag{1}$$

where u represents the perturbed forward speed, w the perturbed heave velocity, q the perturbed pitch rate, and θ, the perturbed pitch attitdue. X_x, Z_x and M_x represent the corresponding dimensional stability derivatives. U_o represents the trimmed speed of the aircraft. δ_{TH} and δ_E represent changes in the thrust delivered by the engines and the elevator deflection, respectively. These two control inputs are what is customarily provided on conventional aircraft; on CCV there are many control surfaces, which will be denoted in this paper by δ_x where x will indicate the appropriate surface. For example, δ_F will denote a change in the flap deflection. Because of the restrictions of space only longitudinal motion will be dealt with in this paper.

The equations in (1) can be represented as a state equation by defining the state vector, x, as:-

$$x' = [u \quad w \quad q \quad \theta] \tag{2}$$

and the control vector, u, as:

$$u' = [\delta_E \quad \delta_{TH}] \tag{3}$$

The corresponding matrices, A and B, are defined as:-

$$A = \begin{bmatrix} X_u & X_w & 0 & -g\cos\gamma_o \\ Z_u & Z_w & U_o & 0 \\ \widetilde{M}_u & \widetilde{M}_w & \widetilde{M}_q & 0 \\ 0 & 0 & 1 & 0 \end{bmatrix} \qquad (4)$$

and

$$B = \begin{bmatrix} 0 & X_{\delta_{TH}} \\ Z_{\delta_E} & Z_{\delta_{TH}} \\ \widetilde{M}_{\delta_E} & \widetilde{M}_{\delta_{TH}} \\ 0 & 0 \end{bmatrix} \qquad (5)$$

where

$$\begin{aligned} \widetilde{M}_u &= M_u + M_{\dot{w}} \cdot Z_u \\ \widetilde{M}_w &= M_w + M_{\dot{w}} \cdot Z_w \\ \widetilde{M}_q &= M_q + M_{\dot{w}} \cdot U_o \\ \widetilde{M}_{\delta_E} &= M_{\delta_{TH}} + M_{\dot{w}} \cdot Z_{\delta_{TH}} \end{aligned} \qquad (6)$$

The change of acceleration experience at the aircrafts' c.g. is given by:-

$$a_{z_{c.g.}} = \dot{w} - U_o q \qquad (7)$$

At any other station, A, on the aircraft at some distance, ℓ_x from the c.g., the acceleration will be

$$a_{z_A} = a_{z_{c.g.}} - \ell_x \dot{q} \qquad (8)$$

ℓ_x is measured positive forward of the c.g. The change of height from the aircraft's trimmed position is related to the normal acceleration as follows:-

$$\ddot{h}_{c.g.} = -a_{z_{c.g.}} \qquad (9)$$

$$\therefore \dot{h}_{c.g.} = -w + U_o \theta \qquad (10)$$

However, the aircraft's flight path angle is defined as

$$\gamma = \theta - \alpha = \theta - \frac{w}{U_o} \qquad (11)$$

$$\therefore a_{z_{c.g.}} = -U_o \dot{\gamma} \qquad (12)$$

The acceleration can be regarded as an output variable defined in terms of the aircraft's state vector and control inputs as follows:-

$$y \overset{\Delta}{=} a_{z_{c.g.}} = Z_u u + Z_w w + U_o q + Z_{\delta_E} \delta_E + Z_{\delta_{TH}} \delta_{TH} - U_o q$$

$$= Z_u u + Z_w w + Z_{\delta_E} \delta_E + Z_{\delta_{TH}} \delta_{TH}$$

$$= [Z_u \quad Z_w \quad 0 \quad 0] x + [Z_{\delta_E} \quad Z_{\delta_{TH}}] u$$

$$= Cx + Du \tag{13}$$

2.2   The dynamics of a flexible aircraft

When aeroelastic effects have to be taken into account, it is necessary to augment the aircraft's rigid body equations by adding to the state variables a set of generalized co-ordinates, associated with the normal bending modes. These modes have been determined by assuming, first, that the structural behaviour is linear and, next, that any structural displacement is small compared to the overall dimensions of the aircraft. The bending equations really describe the aircraft's motion relative to a mean-axes co-ordinate system: with small displacements the stability axes may be assumed to coincide with the mean axes. The bending modes are the normal modes of vibration, in vacuo. With such assumptions, each bending mode can be characterised by a distinct natural frequency, ω_i, and by a vector, v_i. If the ith bending mode, say, is considered to be damped, it can be represented by:-

$$A_i \ddot{q}_i + B_i \dot{q}_i + C_i q_i = Q_i \tag{14}$$

Q_i is a generalized force. A_i, B_i, and C_i are the coefficients of the ith generalized co-ordinate, q_i, and of its corresponding rates.

$$\begin{aligned} Let \quad & x_1 \overset{\Delta}{=} q_i \\ and \quad & x_2 \overset{\Delta}{=} \dot{q}_i = \dot{x}_1 \end{aligned} \tag{15}$$

From equations (14) and (15) it can easily be shown that

$$\begin{aligned} \dot{x}_1 &= x_2 \\ \dot{x}_2 &= -\frac{B_i}{A_i} x_2 - \frac{C_i}{A_i} x_1 + \frac{1}{A_i} Q_i \end{aligned} \tag{16}$$

or, more usually,

$$\dot{x}_2 = -2\xi_i w_i x_2 - \omega_i^2 x_1 + a_i Q_i \dots$$

The ith bending mode has now been represented by two, first order linear differential equations, equation (16). In such a fashion it is possible to augment the rigid-body dynamics with pairs of such first-order equations corresponding to each bending mode being considered. Usually only sufficient generalized coordinates, q_i, are included to represent the aeroelastic effects adequately. If a number of bending modes are to be included in the model, it is the convention to regard as mode 1 that mode with the lowest bending frequency. The mode number goes in ascending order as the frequency associated with each mode increases. In some applications the shortest period associated with the entire motion is long compared with longest period of vibration, $2\pi/\omega$. When that occurs the inertia $(\ddot{q}_i)$ and damping $(\dot{q})$ terms may be negligible. With modern aircraft, however, it is a trend that the shortest period is not usually long compared with the largest period of vibration. Consequently, the inertia terms must be included. For AFCS designers, the question then arises of how many structural bending modes need to be considered to represent adequately the aerolastic effects. A number of methods of determining these are used, but the most common is <u>modal truncation</u>.

Suppose that, for rigid-body, short period motion, the flexibility effects of a conventional fighter aircraft are considered to be adequately represented, using modal truncation, by five bending modes viz 1, 5, 7, 8 and 12. The state vector may now be defined as:

$$\mathbf{x}' \triangleq [\alpha \quad q \quad \lambda_1 \quad \dot{\lambda}_1 \quad \lambda_5 \quad \dot{\lambda}_5 \quad \lambda_7 \quad \dot{\lambda}_7 \quad \lambda_8 \quad \dot{\lambda}_8 \quad \lambda_{12} \quad \dot{\lambda}_{12}] \tag{17}$$

The corresponding matrices, A and B, are shown as Figure 2. The form of the coefficient matrix can be seen to be:-

$$\begin{bmatrix} \textit{Rigid body terms} & \vdots & \textit{Aeroelastic coupling terms} \\ -\,-\,-\,-\,-\,-\,-\,-\, & \vdots & -\,-\,-\,-\,-\,-\,-\,- \\ \textit{Rigid body coupling terms} & \vdots & \textit{Structural flexibility terms} \end{bmatrix}$$

Note that there is coupling between bending modes 7 and 8.

When bending effects are present in the the aircraft dynamics, the acceleration which occurs as a result of structural motion has to be added so that the normal acceleration at station A becomes:-

$$a_{z_A} = \dot{\alpha} - q - \ell_x \dot{q} + \phi_{x,1}\ddot{\lambda}_1 + \phi_{x,5}\ddot{\lambda}_5 + \phi_{x,7}\ddot{\lambda}_7$$

$$+ \phi_{x,8}\ddot{\lambda}_8 + \phi_{x,12}\ddot{\lambda}_{12} \tag{17}$$

The coefficients $\phi_{x,i}$ are the displacement coefficients of the bending modes, which are obtained from graphs provided by the aircraft manufacturer of bending mode deflection versus fuselage or wing station.

In a similar fashion the signals produced from gyros used as feedback sensors are also affected by bending motion. If a vertical gyroscope is used to measure the local inclination at some point A the signal is proportional to:

$$\theta_A = \theta + \sum_j \phi_{Aj} \lambda_j \tag{18}$$

A rate gyroscope, located at the same point, would measure q_A i.e.

$$q_A = q + \sum_j \phi_{Aj} \dot{\lambda}_j \tag{19}$$

2.3   Bending moments

A bending moment of a wing or fuselage is defined by:-

$$V_{B.M._A} = \left\{ EI\left(\frac{-d^2 \xi_A}{dy^2} \right) \right\}_{y=0} \tag{20}$$

where, according to normal mode theory, the deflection at point A can be expressed as:-

$$-\xi_A = \phi_A h + \phi_{12}\theta + \sum_{k>2} \phi_{Ak} \lambda_{k-2} \tag{21}$$

h is found from the relationship:-

$$\dot{h} = U_o \theta - w \tag{22}$$

It can be shown that:-

$$\frac{-d^2 \xi_A}{dy^2} = \frac{d^2 \phi_{A1}}{dy^2} h + \frac{d^2 \phi_{A2}}{dy^2} \theta + \frac{d^2 \phi_{A3}}{dy^2} \lambda_1$$

$$+ \frac{d^2 \phi_{A,4}}{dy^2} \lambda_2 + \dots \tag{23}$$

For rigid-body modes

$$\frac{d^2 \phi_{A1}}{dy^2} = \frac{d^2 \phi_{A2}}{dy^2} = 0 \tag{24}$$

hence

$$V_{B.M._A} = \left\{ EI\left[\frac{d^2 \phi_{A3}}{dy^2} \lambda_1 + \frac{d^2 \phi_{A4}}{dy^2} \lambda_2 + \dots \right] \right\}$$

$$= M_{A1} \lambda_1 + M_{A2} \lambda_2 \tag{25}$$

If the bending moment at some particular wing station, j, is considered to be an output variable, y, say, then where

$$y = C\mathbf{x}$$ (26)

if

$$C \triangleq [0 \quad 0 \quad M_{j1} \quad M_{j2} \cdots \cdots]$$ (27)

$$\mathbf{x}' \triangleq [h \quad \theta \quad \lambda_1 \quad \lambda_2 \cdots \cdots \lambda_n]$$ (28)

3.   ATMOSPHERIC TURBULENCE

3.1   The nature of turbulence

The air through which an aircraft flies is never still. Whenever an aircraft flies in such disturbed air, its motion is erratic. The nature of those disturbances is influenced by several factors, but it is customary to regard turbulence, which occurs above that region of space where the atmosphere behaves as a boundary layer, as belonging to either of the following classes:-

a.   convective turbulence, which occurs in and around clouds. This class incudes thunderstorms.

b.   clear air turbulence (CAT). Below the cloud base the air is heated by direct convection, which causes motion which, together with the many small eddies arising as a result of surface heating, are often regarded as mild CAT. Above a cluster of cumulus clouds a regular, short, broken motion can persist. Violent CAT is usually to be found near mountains, but the most violent turbulence of all is causd by thunderstorms and squall lines, especially when the same area is being subjected at the same time to rain, hail, sleet or snow.

Such atmospheric turbulence is often referred to as atmospheric gusts. Because the mechanisms of turbulence are so varied and complicated, it has been found that the only effective methods of analysing dynamic problems in which turbulence is involved are statistical methods. The mathematical model to be presented next is not entirely descriptive of the phenomenon which it is meant to represent, but it does represent the significant characteristics sufficiently well to permit an analysis to be carried out with adequate accuracy for engineering purposes.

3.2   Continuous gust representation

Two particular representations for the PSD functions of atmospheric turbulence find extensive use in AFCS studies. The Von Karman function is the better fit to the spectrum obtained from records of gusts. It is the less favoured in analytical studies because it is a more complicated PSD, its definition being:-

$$\phi_{VK}(\Omega) = \frac{\sigma^2 L}{\pi} \frac{\left\{1 + \frac{8}{3}(1.339 L\Omega)^2\right\}}{\left\{1 + (1.339 L\Omega)^2\right\}^{\frac{11}{6}}} \tag{29}$$

Because of the non-integer index, the Von Karman PSD function is difficult to simulate directly. The second PSD function, the Dryden model, is the more favoured form because it is simpler and more easily programmed:

$$\phi_{DRY}(\Omega) = \frac{\sigma^2 L}{\pi} \frac{(1 + 3L^2\Omega^2)}{(1 + L^2\Omega^2)^2} \tag{30}$$

The difference between these functions is not great, so that when a small fighter aircraft is being considered, for example, when $0.1 \le \omega < 8.0$, the two forms are within a few dB of each other.

Since

$$\omega = U_o \Omega \tag{31}$$

then

$$\phi_{DRY}(\omega) = \frac{\sigma^2 L}{\pi} \frac{\left(1 + \frac{3L^2}{U_o^2}\omega^2\right)}{\left(1 + \frac{L^2}{U_o^2}\omega^2\right)^2} \tag{32}$$

L is the scale length of the turbulence. To generate a Dryden gust signal with the required gust intensity, scale length, and PSD function, a wide band noise source with a PSD function $\phi_N(\omega)$, is used as an input signal to a linear filter-chosen so that it has an appropriate frequency response such that the output signal from the filter will have a PSD function, $\phi(\omega)$, very nearly the same as that defined for the Dryden model.

3.3   State variable representation of atmospheric turbulence

The Dryden model of vertical gust velocity is given by:-

$$w_g(s) = \sqrt{K}\,\frac{(s + \beta)}{(s + \lambda)^2}\,\eta(s) \tag{33}$$

where $\eta(t)$ is a white noise signal,

$$K = \frac{3U_o \sigma_w^2}{L_w \pi}$$

$$\beta = \frac{U_o}{\sqrt{3L_w}} \tag{34}$$

and

$$\lambda = U_o/L_w$$

It can be shown easily that if the output variable, y_g, is defined as the vertical gust velocity, w_g, then an appropriate state variable model is:-

$$\dot{\mathbf{x}} = A_g\mathbf{x} + B_g\eta \tag{35}$$
$$y_g = C_g\mathbf{x}$$

where

$$A = \begin{bmatrix} 0 & 1 \\ -2\lambda & -\lambda^2 \end{bmatrix} \tag{36}$$

$$B' = [0 \qquad 1] \tag{37}$$

and

$$C = \sqrt{K}\,[1 \qquad \beta] \tag{38}$$

The gust terms are introduced into the aircraft state equation by addition i.e.

$$\dot{\mathbf{x}} = A\mathbf{x} + B\mathbf{u} + Ey_g \tag{39}$$

4.   ACT SYSTEMS

4.1   Stability augmentation system

From the matrix, A, shown in Figure 2 it can easily be determined that the rigid-body, short-period dynamics of a modern fighter, which are chiefly characterized by the angle of attack, α, and the pitch rate, q, are governed by the sub-matrix:-

$$\begin{bmatrix} Z_\alpha & 1 \\ M_\alpha & M_q \end{bmatrix}$$

The corresponding characteristic polynomial is given by:-

$$\lambda^2 - (Z_\alpha + M_q)\lambda + (-M_\alpha + Z_\alpha M_q)$$

Obviously, the damping ratio of the short period dynamics is given by:-

$$\zeta_{sp} = \frac{-(Z_\alpha + M_q)}{2\sqrt{-M_\alpha + Z_\alpha M_q}} \tag{40}$$

To augment the dynamic stability requires that either Z_α or M_q (or both) be increased by means of feedback control. Note that the value of the static stability derivative, M_α, is always negative for figher aircraft possessing static stability.

Making M_α less negative, even positive, and thereby relaxing the aircraft's static stability, will also increase the short period damping somewhat but its principal effect will be to improve the fighter's manoeuvrability. However, it is not easy to change M_α by feedback control. Moreover, the aircraft time constant, T_A, is inversely proportional to Z_α; if T_A is too great, in relation to ζ_{sp} and w_{sp}, the resulting peak overshoot in the aircraft's indicial response will be unacceptable. Consequently, it is desirable to augment Z_α, but, in general, this can only be done effectively on CCV where direct lift controls, such as canards, are available. The principal means of achieving stability augmentation is to use a negative feedback signal proportional to pitch rate as the control signal to the aircraft's elevator. The response of a typical, single-engined, CCV fighter of the type represented in Figure 1, to an initial disturbance in its angle of attack when in supersonic flight, is shown as curve A in Figure 3. The corresponding response for a SAS fitted to the same aircraft, and based on the feedback control law

$$\delta_E = \quad 0.5q \tag{41}$$

is shown as curve B in the same figure. The same response can be obtained from a SAS which involves the use of the horizontal canards in the feedback control law i.e.

$$\delta_E = .2\alpha + 0.5q \tag{42}$$

$$\delta_{CAN} = -.022\alpha - .038q$$

4.2   Gust load alleviation (GLA) and bending mode control (BMC)

4.2.1   Gust alleviation

The amplitude of the response caused by structural vibration excited by turbulence may be reduced if:-

(a)   the amount of energy transferred from the gust to the bending modes is reduced;

or (b)   any energy which has been absorbed by those modes is rapidly dissipated.

For effectiveness, both methods are usully employed simultaneously. To reduce the energy being transferred requires that a countering moment (or force) be generated by deflecting some control surfaces. However, the transfer reduction method requires an accurate knowledge of the aircraft's stability derivatives, but such derivatives change with many factors such as the aircraft's flight condition, with changes in its mass, with changes in dynamic pressure etc. As a result, the aircraft dynamics will be known too imperfectly to admit of perfect cancellation of any gust forces or moments. Once energy has been absorbed, however, its dissipation can be controlled by augmenting the damping of the elastic modes. If the modes are too close in frequency, however, it is difficult to achieve a sufficient increase in structural damping for they are usually then too closely-coupled and there occurs a periodic exchange of energy between the modes, behaviour which corresponds to a very lightly-damped structure. For active suppression of structural bending it is essential to be able to sense the structural displacements and their corresponding rates of change to provide the feedback signals required for a GLA system.

4.2.2   GLA and BMC

Both systems perform the same function: the reductions of loads imposed on the aircraft as a result of its motion. For GLA the loads are usually the accelerations experienced by the aircraft as a result of flying in gusty conditions. BMC is intended to reduce the bending moments of the wing or fuselage (often both) which arise as a result of any aircraft motion. From the control viewpoint they can both be regarded as output regulator problems and one method of obtaining an appropriate feedback control law is to solve the linear quadratic output regulator problem where the performance index, J, is minimized.

$$J \overset{\Delta}{=} \tfrac{1}{2} \int_0^\infty (\mathbf{y}'Q\mathbf{y} + \mathbf{u}'G\mathbf{u})\, dt \tag{43}$$

For GLA the output vector will be of the type shown as equation (13), for example. For BMC, the appropriate output vector will be that corresponding to eqs (26) – (28).

Figure 4a shows the normal acceleration responses of a modern fighter aircraft subjected to a Dryden gust with zero mean value and of 0.33 m/s standard deviation. The responses are measured at the pilot's station (NZ172) and the fighter's c.g. (NZ860). This fighter has a number of significant bending modes which accounts for the high frequency oscillations excited by the turbulence.   The A curves represent when the aircraft has no GLA control system; the B curves show the responses when the GLA system is operating.   In Figure 4b the vertical bending moments at the same aircraft fuselage stations are shown for the same fighter aircraft flying at the same speed and height through the same gust.   Note how effective the BMC system has been.

Both the GLA and BMC systems involved the use of the horizontal canard control surface in addition to the use of the elevator.

<u>4.3   Ride control systems</u>

<u>4.3.1   Ride quality</u>

Almost every modern fighter aircraft has a SAS to control its rigid body motion. In such SAS the locations of the feedback sensors are carefully chosen to result in the minimum amount of spurious signals from any structural motion being picked up.   Such SAS are not designed to control or deliberately to alter the structural vibration of the aircraft, although such SAS do provide by their action a large reduction of the unwanted motion produced by the aircraft in response to a gust. From studies of operational records and simulations it is known that it is those symmetrical structural modes with the lowest frequencies which contribute most to the levels of acceleration present at various points of the fuselage, such as the cockpit. For example, at the cockpit of a typical modern fighter aircraft, it has been found that, when its SAS is not operating, 60% of the normal acceleration being measured could be attributed to the first three bending modes; of the remaining 40% three quarters was due to rigid-body motion and the other quarter was caused by higher frequency bending modes. If such accelerations are unacceptable, resulting in discomfort to the pilot or impairing his ability to fly the particular mission being undertaken, there is then a need for a ride control system to reduce these levels of acceleration.

<u>4.3.2   Ride discomfort index</u>

A ride discomfort index has been proposed in the military specification document MIL-F-9490D. (ref. 1). It is proportional to

$$\frac{wing\ lift\ slope}{wing\ loading}$$

i.e.

$$J_{RD} = \frac{kC_{L_\alpha}}{W/S} \tag{44}$$

where W is the weight of the aircraft, S is the area of the lifting surface and C_{L_α} is the slope of the lift characteristic.

For any aircraft, for which the equations of motion are expressed in the stability axis system, the normal force stability derivative with angle of attack, Z_α, can be shown to be:-

$$Z_\alpha \simeq -\frac{\rho S U_o}{2m}C_{L_\alpha} \tag{45}$$

where m is the aircraft's mass, ρ is the relative density of the atmosphere, and U_o is the aircraft's equilibrium speed

$$\therefore \qquad Z_\alpha = -\frac{\rho U_o g}{2k} J_{RD} \qquad (46)$$

g is the acceleration due to gravity. However, it has already been shown that the normal acceleration can be expressed approximately as

$$a_z = Z_\alpha \alpha + \sum_{j=1}^{m} Z_{\delta_j} \delta_j$$

$$= -\frac{\rho U_o g}{2k} J_{RD} \alpha + \sum_{j=1}^{m} Z_{\delta_j} \delta_j \qquad (47)$$

Thus, for any given control activity, if J_{RD} is minimized, the corresponding normal acceleration will also be minimized.

4.3.3   Determination of r.m.s. acceleration caused by a gust

To assess the ride control problem it is necessary to know the r.m.s. values of the acceleration being measured at a particular aircraft location. Although these values can be found using power spectral density functions, a more efficient method for computation is given in ref. (2).

The gust signal, y_g, can be obtained from

$$\dot{y}_g = F y_g + D \eta \qquad (48)$$

Combining eq. (39) and (48) yields:-

$$\begin{bmatrix} \dot{x} \\ \dot{y}_g \end{bmatrix} = \begin{bmatrix} A & E \\ 0 & F \end{bmatrix} \begin{bmatrix} x \\ y_g \end{bmatrix} + \begin{bmatrix} B \\ 0 \end{bmatrix} u + \begin{bmatrix} 0 \\ D \end{bmatrix} \eta \qquad (49)$$

The response of the aircraft to any control inputs is ignored. If it is important to consider the response to both gusts and inputs simultaneously the responses can be found independently and summed, since the mathematical model representing the aircraft dynamics is linear.

By defining

$$x^* = \begin{bmatrix} x \\ y_g \end{bmatrix} \qquad (50)$$

then eq. (49) can be expressed as

$$\dot{x}^* = H x^* + M \eta \qquad (51)$$

where

$$H = \begin{bmatrix} A & E \\ 0 & F \end{bmatrix} \qquad (52)$$

and

$$M = \begin{bmatrix} 0 \\ D \end{bmatrix} \tag{53}$$

When only gusts are being considered the normal acceleration $a_{z_{cg}}$, can be represented by

$$y \overset{\Delta}{=} a_{z_{cg}} = C\mathbf{x}^* \tag{54}$$

The mean value of $a^2{}_{z_{cg}}$ can be found by squaring $a_{z_{cg}}$ and averaging i.e.

$$a^2_{z_{cg}} = \begin{bmatrix} C\mathbf{x}^* \end{bmatrix}\begin{bmatrix} C\mathbf{x}^* \end{bmatrix}'$$

$$= C\mathbf{x}^*x^{*\prime}C' \tag{55}$$

The mean value is the expected value of $a^2{}_{z_{cg}}$

$$E\begin{bmatrix} a^2_{z_{cg}} \end{bmatrix} = C\,E\begin{bmatrix} \mathbf{x}^*x^{*\prime} \end{bmatrix}C' \tag{56}$$

where $E[\mathbf{x}^*x^{*\prime}]$ is the state covariance matrix of order $(n+1) \times (n+1)$.

It can be shown that:

$$H\,E[\mathbf{x}^*x^{*\prime}] + E[\mathbf{x}^*x^{*\prime}]\,H' + MM' = 0 \tag{57}$$

When H and M are known there is a unique solution for $E[\mathbf{x}^*x^{*\prime}]$

Eq. (57) is in the form

$$BX' + XB' = -C \tag{58}$$

and there are many algorithms available for solving the Lyapunov equation

$$XA + A'X = -Q \tag{59}$$

To use any of these algorithms it is only necessary to arrange that the following are true:-

$$Q = C = MM'$$
$$A = H \tag{60}$$

Using these values of Q and A in eq. (59) will result in a solution for $E[\mathbf{x}^*x^{*\prime}]$, and hence, by means of eq. (56), of the rms acceleration.

4.3.4   Ride control system

With a rate gyro, measuring pitch rate, and an accelerometer, measuring the normal acceleration at the fighter's cockpit, providing the feedback signals, the RCS uses elevator and horizontal canards as the control surfaces. The feedback control law can most easily be obtained as the solution of a linear optimal quadratic regulator problem in which a performance index, of the type given as eq. (43), with

4

the output vector, y, being defined by eq. (47), is minimized. The optimal feedback control law will reduce the acceleration and the ride discomfort being experienced by the pilot. If the standard deviation of the Dryden gust is adjusted such that the r.m.s. value of acceleration experienced by the pilot of the figher aircraft with the RCS in operation is unity, then, using the method outlined in section 4.3.3, it can be shown that with the RCS the r.m.s. acceleration for the same gust intensity can be reduced to less than 0.1.

5. CONCLUSION

Many of the ACT systems discussed in this paper can be found in modern fighter aircraft. Most have been designed using modern control methods, such as the optimal linear quadratic regulator problem, to determine the appropriate feedback control law. Other methods such as model-following or eigenstructure assignment can be used with equal facility, but the resulting control law will take the same form namely a linear full, state-variable feedback control law. When GLA or BMC is being considered, where the most significant bending modes are taken into account, the corresponding feedbck control will require that the bending displacements and rates are available for measurement. This may be unlikely, and if those methods are to be employed, then recourse to state estimation techniques will be required.

6. REFERENCES

1. Anon
   Military Specifications - Flying Qualities of Figher Aircraft.
   MIL-F-9490D, Dept. of Defense, U.S.A.

2. Swain, R.L., D.K. Schmidt, P.A. Roberts & A.J. Hinsdale
   An Analytical Method for Ride Quality of Flexible Airplanes
   AIAA Journal, 1977, 15(1), 4-7.

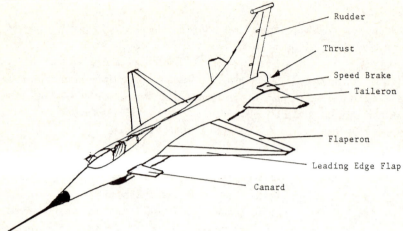

- Rudder
- Thrust
- Speed Brake
- Taileron
- Flaperon
- Leading Edge Flap
- Canard

Figure 1

$$A = \begin{bmatrix}
Z_a & 1 & s & t & 0 & 0 & 0 & 0 & 0 & 0 & 0 & 0 \\
M_a & M_q & u & v & 0 & 0 & 0 & 0 & 0 & 0 & 0 & 0 \\
0 & 0 & 0 & 1 & 0 & 0 & 0 & 0 & 0 & 0 & 0 & 0 \\
a & b & -\omega_1^2 & -2\xi_1\omega_1 & 0 & 0 & 0 & 0 & 0 & 0 & 0 & 0 \\
0 & 0 & 0 & 0 & 0 & 1 & 0 & 0 & 0 & 0 & 0 & 0 \\
c & d & 0 & 0 & -\omega_5^2 & -2\xi_5\omega_5 & 0 & 0 & 0 & 0 & 0 & 0 \\
0 & 0 & 0 & 0 & 0 & 0 & 0 & 1 & 0 & 0 & 0 & 0 \\
e & f & 0 & 0 & 0 & 0 & -\omega_7^2 & -2\xi_7\omega_7 & w & x & 0 & 0 \\
0 & 0 & 0 & 0 & 0 & 0 & 0 & 0 & 0 & 1 & 0 & 0 \\
g & h & 0 & 0 & 0 & 0 & y & z & -\omega_8^2 & -2\xi_8\omega_8 & 0 & 0 \\
0 & 0 & 0 & 0 & 0 & 0 & 0 & 0 & 0 & 0 & 0 & 1 \\
k & \ell & 0 & 0 & 0 & 0 & 0 & 0 & 0 & 0 & -\omega_{12}^2 & -2\xi_{12}\omega_{12}
\end{bmatrix}$$

$$B' = \begin{bmatrix}
Z_{\delta_E} & M_{\delta_E} & 0 & V_{11} & 0 & V_{21} & 0 & V_{31} & 0 & V_{41} & 0 & V_{51} & 0 \\
Z_{\delta_{TH}} & M_{\delta_{TH}} & 0 & V_{21} & 0 & V_{22} & 0 & V_{32} & 0 & V_{42} & 0 & V_{52} & 0
\end{bmatrix}$$

Figure 2

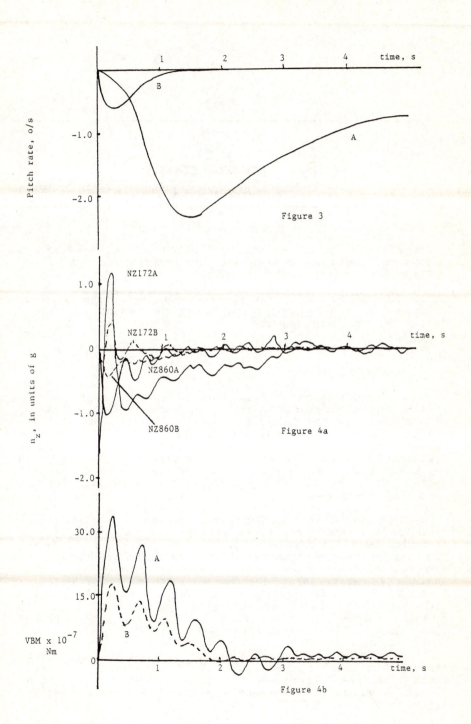

Figure 3

Figure 4a

Figure 4b

Chapter 16

Batch process control

J. D. F. Wilkie

Some of the characteristics of batch control systems are discussed, and illustrated with particular reference to an application from the sugar industry.

A comparison of experience with different control equipment is touched on, and related to present and future directions for development.

1.   INTRODUCTION

The control of batch processes is an important area of application of computer control systems. The installation of computer systems has often been justified more for the ability to handle recipes and maintain batch records than for any complexity of control. A batch process is indeed typified by short runs of different products, though there may also be multiple streams, and the ability of a computer system to keep track of different batches is another important factor.

In a multi-product environment the major benefit arises from the assurance that waste will be reduced when changing from one batch to another. In the pharmaceutical industry the legislative requirements demand that precise records of each batch are maintained. Computer control not only assists in maintaining the records, it makes a positive contribution to the quality assurance for which the records are required. This is not simply the application of quality control and the rejection of bad batches, but the assurance that comes from eliminating the bad batches.

Whichever of these varied justifications is used, there are characteristics which are common to almost all batch control systems. Examining an application in the beet sugar industry should illustrate aspects of more general relevance.

2.   BATCH PROCESSES IN THE BEET SUGAR INDUSTRY

The entire beet sugar industry could be regarded as a batch process, in that it follows the agricultural cycle and operates for a campaign period of typically 120 days from September to January each year. However, the problems of start-up and shut-down in these circumstances, although allied to those of batch process control, do have a number of significant differences. The characteristics of the raw material (the beet) vary from year to year with the weather; new equipment may have been installed in the factory to enhance capacity or efficiency. Taken with the lengthy off-season and changes in staff, there is a regular need for operators and management to re-learn the detailed refinements of the process operation. It is the "once-a-year" nature of the problem which makes it different. The need to get from an off condition to a running condition, or vice versa, as quickly and efficiently as possible, is at the nub of much batch process control.

Once the campaign is underway the sugar production process is essentially continuous although there are a number of sub-processes where mechanical limitations involve start-up and shut-down during the campaign. From a control viewpoint these could be regarded as batch processes. However, there is one aspect of the sugar process which remains predominantly batch, and which is of fundamental importance. This is the crystallisation process which is carried out under vacuum in vacuum pans of typically 70 to 90 tonnes capacity. In order to achieve the correct purification of the final product this is a multi-boiling process in which there is successive crystallisation and centrifugal separation with the mother liquor from each separation being re-crystallised and these lower purity crystals being re-dissolved and recycled to the earlier stages. The vacuum pans at each stage are known as "white", "raw", and "after product", (or AP).

The crystallisation process is economically important because the two resulting products, white sugar and molasses, have markedly different values. The entire process is effectively limited by the raw material supply (the quantity of sugar beet grown by the farmers) so there are clear advantages in extracting more sugar. At the same time the energy demands of the separation process need to be minimised.

3.   THE VACUUM CRYSTALLISATION OF SUGAR

A vacuum pan is essentially a single-effect evaporator, specially designed for the application. Low pressure exhaust steam, (or vapours), from previous evaporation stages are used to evaporate water. Operation under vacuum allows the water to be evaporated at a lower temperature, thereby reducing the quality of vapour required while at the same time minimising the formation of colour in the crystals.

The basic requirement of the vacuum pan operation is to take a solution containing sugar and non-sugars and to produce crystals of a specified size and shape.

The composition of the feed to the vacuum pan is characterised by its brix and its purity. Brix is a comparison of the mass of total dissolved solids to the total mass of solution expressed as a percentage. Purity is a similar measure of the amount of sucrose in the total dissolved solids. The brix and purity will affect the consistency, or viscosity, of the solution. More importantly they affect the solubility of the sucrose at a given temperature and thus the amount of sucrose required to have a "saturated" solution. For crystallisation to occur the solution must be super-saturated. The precise effects of purity on saturation depend on the particular impurities, and will be affected by the nature of the land where the beet has been grown, and often vary as the campaign progresses.

A typical vacuum pan and the associated instrumentation and valves are shown schematically in Figure 1, which is based on an AP pan. The feed tanks, the strike receiver and the condenser are all common to a number of vacuum pans. The strike receiver is so called because each individual batch is known as a strike.

3.1   The operations of a strike

The instrumentation shown in Figure 1 is best explained in relation to the cycle of operations in a strike. This can be summarised as :

Raise Vacuum

Charge

Boil down

Seed

Feed

Brix up

Discharge and steam out

3.2   Raise Vacuum

In the first phase the vacuum is established in the pan, in a manner which minimises the effect on other pans, and allows the vacuum to be controlled once it is established. This generally involves a separate small valve which bypasses the control valve and is opened until the vacuum comes within the control band.

3.3   Charge

In the second phase the pan is charged with standard liquor (or with a mixture of one or two syrups for raw and AP pans), generally until the level is above the calandria into which the steam is introduced.

Where two syrups are used, the ratio is determined to give a target purity of the charge, and the levels may be adjusted by the operator as the purity of the feed syrups varies. The seed pot is cleared during the charge phase, and the stirrer started at an appropriate level.

3.4   Boil Down

The syrup is then boiled down by increasing the steam flow into the calandria and increasing the density and supersaturation of the syrup. The level in the pan is maintained by feeding syrup, until approaching the critical supersaturation condition when the feed valve is closed. This critical supersaturation is determined by one of several measurements :-

(a)   Syrup Temperature. This is a measure of the boiling point elevation - the temperature to which the sucrose solution has to be raised for boiling to occur above that of water at the same absolute pressure. It is a simple and low cost measurement, but dependent on good control of the absolute pressure and since it is sensitive to variations with purity it is generally only used on high purity (white) pans. An accurate measurement is required, since a temperature change of 1.35°C can change supersaturation from 1.0 to 1.2; from redissolving sugar to conditions for spontaneous nucleation.

(b)   Consistency/Viscosity. Some form of paddle is inserted in the wall of the vacuum pan and the resistance to motion is converted into an analogue signal. This measurement has been widely used in continental Europe but is prone to unreliability.

(c)   Density. Either using an optical refractometer or a radiometric density gauge. The density measurement is tending to replace the consistency "rheometer" in Europe because it requires little maintenance. While the optical device will give the same measurement so long as all the sugar is in solution, as soon as crystallisation starts to occur, the optical device responds to the density of the "mother liquor" surrounding the crystals. The Finnish Sugar Company has developed a control system which uses both measurements in order to determine the quantity of sucrose that has crystallised out [4].

(d)   Electrical Conductivity. This has been found to be a simple, low cost indication of the extent of supersaturation. However, it is not a precise measure, and the presence of sugar crystals in the solution affects the electrical path length and thus the conductivity. It is widely used particularly on low purity (AP) pans. Several instruments have been developed which measure capacitive effects at a range of frequencies but all are subject to the same ill-defined effects of crystal content.

Whichever measurements are used, the positioning of the primary element in the pan has to be chosen to ensure a representative signal is obtained. It must not be influenced by incoming feed, poor circulation or vapour bubbles.

3.5   Seed

Once the correct condition has been reached the pan is seeded. Generally this is done by introducing a fixed quantity of milled seed crystals in a suspension of isopropyl alcohol. The sugar juice then begins to crystallise around these seed crystals, causing them to grow in their fixed shape. This suspension of sugar crystals in a surrounding syrup is known as massecuite. It is also possible to "shock seed" a pan by a sudden change in vacuum, or the introduction of larger quantities of seed sugar, which increase the supersaturation to the point where spontaneous nucleate crystallisation occurs. Whichever method is used the period following seeding is critical, since the aim is to have a single generation of crystals which will grow at the same rate, and be of consistent size at the end of the batch. If the super- saturation level falls too far, the crystals will dissolve. If it increases too far, there will be further spontaneous nucleation.

To overcome the need for careful control on every pan there is a trend towards using a dedicated seeding system (either a batch pan or a cooling crystalliser) to grow seed from 10 micron to 100 micron. A much larger volume of seed has to be fed to the other pans, but it is less likely to be re-dissolved and so the pan supersaturation can be lowered to reduce the risk of spontaneous nucleation.

3.6   Feed

Once the grain is established, the syrup feed is controlled to increase the pan level and maintain the supersaturation of the mother liquor. All the measurements mentioned above, except the optical refractometer, will be affected by both mother liquor supersaturation and the crystal content. In addition, as the level increases, there will be a hydrostatic head effect at the measurement point.

3.7 Brix-up

When the pan is full there is a period of brixing up in which water is further evaporated with no additional feed. Generally this continues until the viscosity or conductivity reaches a pre-determined value. The steam valve to the calandria is then closed.

3.8 Discharge

Finally the vacuum is broken, the pan discharged to a receiver for centrifuging, and the pan steamed out in readiness for the next cycle. The process sounds straight-forward enough, but a number of problems can arise. These stem particularly from the creation of false grain (unwanted spontaneous nucleation) which often leads to uneven and irregular crystals, and conglomerates (crystals sticking together) which often entrap impurities and make centrifuging more difficult. Yet is is clear that as the level in the pan increases the uniformity within the pan will be markedly affected by the hydrostatic head, and it is difficult to avoid variations in supersaturation. Good circulation becomes important.

4. CONTROL SYSTEMS FOR VACUUM PAN CRYSTALLISATION

Vacuum pan operation has been an obvious candidate for computer control for more than 20 years [5]. After early experiences with a single computer controlling all pans [1], British Sugar adopted a micro- processor-based approach controlling not more than two pans [2]. More recently it has been possible to use commercially available control systems with the same constraint. Although each system is programmed in a slightly different way, the main changes have been in the displays and other features available to the process operators on modern systems. These make it easier to follow how the control system is performing, and in some cases easier to program, but the fundamental control requirements have not altered.

It has been found that programming techniques based on distinct phases with appropriate steps in the sequence have been easier to support than those based on parallel logic. This approach is typical of some PLCs and will work. However it is extremely clumsy and difficult to break into segments which another engineer can understand.

As with other batch control systems there is a mixture of sequence control (charging, seeding, discharging and steaming out) and continuous control (of vacuum, level and steam flow). The continuous control loops are illustrated in Figure 2. Not only is there sequence control on the on/off valves but the continuous control loops themselves are subject to sequentially controlled changes. The vacuum is often ramped to adjust its effect on supersaturation as the strike proceeds.

The feed rate is used to control level directly before seeding, but during feed is used to control boiling point elevation or conductivity, which is itself set as a function of level. It is usually necessary to adjust the loop tuning parameters as the level increases and the loop gain changes. So it is not only set points but tuning parameters and even control regimes that may be altered in the course of the batch.

4.1   Operator Interaction

At present, the sequence is not fully automated. In particular it is necessary for an operator to fill the seed pot on the vacuum pan once the pan has been charged and while it is boiling down. The seed is in suspension in isopropyl alcohol and will tend to come out of suspension if it is left in the seed pot too long. The economics of automating the existing procedures make it difficult to justify, although similar systems have been implemented on the continent. The need to accommodate operator actions is another typical feature of batch control systems.

4.2   Indirect Measurements

Another difficulty associated with the control of a vacuum pan, which is perhaps not typical of batch processes, is the great difficulty of measuring the parameters which we are really trying to control. There is no direct measurement of the supersaturation of the liquor, nor of the crystal content in the pan. Yet during the feeding cycle we would like to adjust the supersaturation to give a maximum growth rate for a particular crystal size. In general, use is made of a target trajectory for the controlled parameter (boiling point elevation or conductivity), and this trajectory itself adjusted if satisfactory results are not being obtained. From time-to-time this open-loop aspect of pan automation has been a hindrance since it is not easy to know how the trajectory should be adjusted to achieve a particular effect.

4.3   Interactive Control

The entire process within the vacuum pan is highly interactive. Although there are only three independent controls (vacuum, steam and feed), changes to any of the three can affect the supersaturation. It should therefore be possible to achieve good crystal quality in a variety of ways. In practice there is a need to maintain throughput (the length of each strike) and to minimise the energy required, and the available steam and achievable vacuum are often limited. These constraints limit the degrees of freedom of any particular control system. Some interesting theoretical work has been done in France in looking at the optimum performance of vacuum pans [3]. However, translating this into reality requires careful analysis of the relationship between the available measurements and the supersaturation and viscosity of the actual syrups in use.

Figure 2 illustrates ideal traces for automatically controlled strikes. Real traces follow this pattern, but are subject to a variety of disturbances.

5. EQUIPMENT FOR BATCH PROCESS CONTROL

As has been said this particular application of batch process control has been implemented on a number of different types of control equipment. The requirements that have been identified - such as the need to adjust control parameters - have limited the ranges of equipment that can be used. This level of flexibility often accompanies a rather basic structure of control system software, which is less easy for a non-expert to understand and modify. Wherever possible use should be made of systems in which the closed-loop control functions are well-defined and easily recognisable structures. This has the added advantage that the associated display functions are familiar. The operator has the confidence that he could take control if required, and is in fact happier to allow the system to operate automatically.

This is perhaps the most significant difference that has been observed between the first micro-processor controllers, which were regarded as black boxes and the most recent systems in which the ability to display target values as well as actual values on faceplates and on graphic displays has helped to obtain a better understanding of the operations both for commissioning and for routine operation.

It seems likely that this experience could be transferred to any other application. Keep the control system as simple as possible; be clear about what special facilities will be required and ensure that they can be achieved without undue complexity (most systems can be moulded to meet specific requirements, but some more easily than others); give careful consideration to operator interactions and try to make the control system behaviour as clear as possible from the displays.

5.1 Control System Integrity

British Sugar does not have a large central engineering staff, and has not been in a position to support a dedicated computer control system specially written for the application. However, even if we had been in such a position, it is doubtful whether this would be a better way to go. The crystallisation process, although important, is only one part of the entire production system, and the control system needs to be co-ordinated with other units in the factory.

Because the vacuum pans operate on a single product it has been possible to maintain our policy of separate controls for a maximum of two pans. However, as the power of distributed control systems has been progressively enhanced it is a policy which needs to be reviewed. A system with dual processors for high reliability and a capacity for handling a larger number of pans may be a better solution. Certainly, if the controls are to be incorporated in a plant-wide scheme then this approach would bring a greater ease-of-use. The systems for which the techniques of control system definition are best developed are those designed for processes which are, of necessity, larger and more complex. To take advantage of these techniques means that a larger number of simpler systems must be handled within one processor.

6.   EXPERIENCE AND BENEFITS

In many instances, British Sugar has moved from totally manual operation of vacuum pans, with chain wheel operation of valves, and visual inspection of the developing grain. Operators obviously appreciate the immediate improvement in their working conditions, and are keen to take advantage of the new technology. The improvements in the first campaign of operation are sometimes limited by plant conditions and have to be followed up in subsequent seasons. Control systems still tend to be less tolerant of plant inadequacies than do humans. Unfortunately humans are almost too flexible and continue to overcome problems when they should complain and ensure that improvements are made. While it is desirable that control systems should develop "intelligence" and continue to perform well over a wider regime, it is important that they can recognise when a problem exists.

There is no doubt that the wider utilisation of these control systems has improved the consistency of the operation. Problems of course remain, the question of measurement of supersaturation and of crystal content. Some thought has been given to the measurement of crystal size distribution in the massecuite. Many pans have microscopes attached which enable operators to view the crystals; and automatic particle size analysers have been considered. However, these are really only effective just after seeding when the crystals are small. At later stages in the strike the crystals are too dense to allow accurate assessment.

There is always scope for improvement. The new technologies of image analysis, optical measurements and expert systems may all be able to play a part in this particular batch control system. However, the widespread benefits of such techniques in batch process control, is more questionable. What is not in dispute is the benefits of computer control for batch processes-in consistency of operation, interaction of control, and information management.

REFERENCES

1. Bass R J, Branch M F, and Donovan J. Computer Control of Sugar Boiling, 22nd British Sugar Technical Conference, May 1974.

2. Bass R J, and Donovan J. Microprocessor Control of Sugar Boiling, 23rd British Sugar Technical Conference, June 1976.

3. Bonnenfant Ph. Application of Numerical Analyses for Pan Automation.   27th British Sugar Technical Conference, June 1984.

4. Virtanen J.   Microprocessor   Control   of   Sugar Crystallisation. International Sugar Journal 1984, Vol 86 No 1026.

5. Knøvl E A and Møller G R.   Progress in Automatic Pan Boiling, Sugar Technology Reviews Elsevier 1975/76.

ACKNOWLEDGEMENTS

Thanks are due to many colleagues in British Sugar who have willingly shared their experiences of this topic.

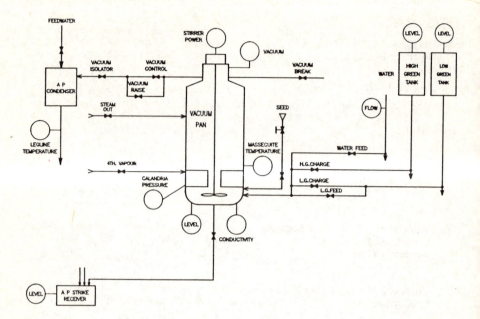

Figure 1.   A P PAN AND INSTRUMENTATION

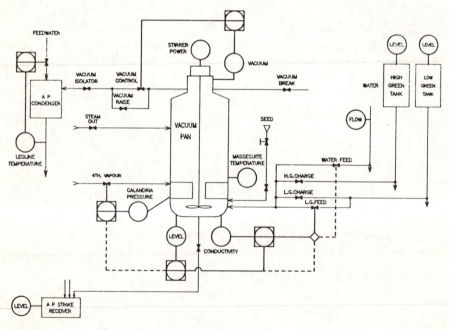

Figure 2.   A P PAN CONTROL LOOPS

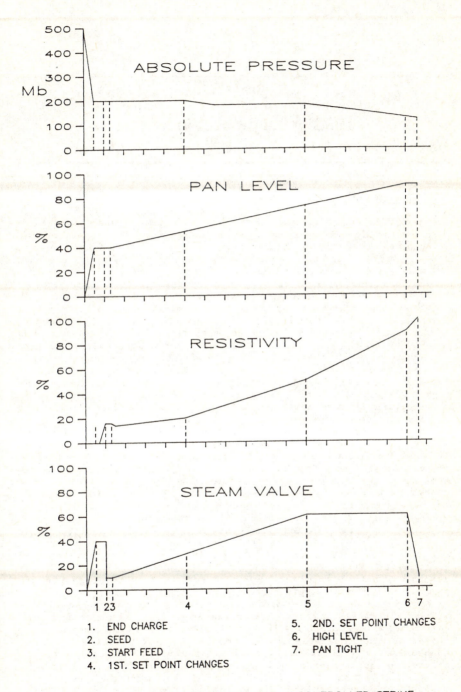

Figure 3.   IDEALISED TRACES FOR A CONTROLLED STRIKE

Chapter 17

DDC in the process industries

D. J. Sandoz

1. INTRODUCTION

The term DDC (Direct Digital Control) is one that is generally used to
describe the digital implementation of conventional analogue control
techniques. This chapter is concerned not solely with DDC but also with
the much broader aspects of online process computing and their relevance
to today's requirements for industrial control systems.

The first use of computers to control industrial processes occurred
in the early 1960's. There is, inevitably, a conflict of claims as to
which country and company made the first application. In Britain, ICI
led the way with an application on a chemical plant at Fleetwood in
Lancashire. A Ferranti Argus 200 computer was used. This computer was
programmed by physically inserting pegs into a plug board, each peg
representing a bit in a computer memory word (it proved more reliable
than its early successors because the destruction of memory contents
required the dislodgement of pegs rather than the disruption of the
magnetic status of ferrite cores). A British computer manufacturer in
league with a British chemical company was amongst the first pioneers of
the application of computers for process control.

The situation has changed. The technology of both hardware and
software has progressed rapidly. The British computer industry has lost
its early initiative and the greater part of progress today is motivated
by developments from the United States. The biggest impact of all has
come from the microprocessor. Computer control systems, once
prohibitively expensive, can now be tailored to fit most industrial
applications on a competitive economic basis. The use of
microprocessors for process control is now the norm.

These advances have motivated many changes in the concepts of the
operations of industrial processes. Video display terminals now provide
the focus for operators to supervise plant. Large panels of
instruments, knobs and switches are replaced by a few keyboards and
screens. Control rooms are now much smaller and fewer people are
required to supervise a plant.

Process computers now have the capability to implement sophisticated
mathematical analysis to aid effective operation. Plant managers and
engineers can be provided with comprehensive information concerning the
status of plant operations. This motivates more effective overall
management of process plant. Surprisingly, and in contrast, the
concepts of basic feedback control implemented by the computers have
changed little from the days when pneumatic instrumentation was the main
means for implementation. Direct Digital Control is essentially a

computer implementation of techniques that have long been established as standard for industrial process control.

There are now signs of movement in this area. A new breed of microprocessor instruments and techniques is emerging. These have the capability for automatically tuning controller parameters for best operating performance. New concepts of logic are evolving for "ruled based" control systems. This expert system approach is intended to replace the need for continuous human involvement in the hour by hour operations of production. Advanced control techniques, such as model based predictive control, are being applied in the Petrochemicals industry and elsewhere for optimisation of process operation. They give rise to high cost benefits in particular situations. Statistical Process Control has emerged as a popular online facility. Today, computer applications of SPC are very simple and generally aid manual procedures, but there is the potential for strong development in this field to achieve complete automation.

This chapter reviews the main functions of a process control computer, including measurement and actuation, direct digital control, sequence control, supervisory control and operator communications. The hardware and software aspects of distributed and hierarchical computer control systems and the integrity and economics of computer operations are discussed. The intention is to provide a broad introductory overview of the significant features of computer control systems.

2.   THE ELEMENTS OF A COMPUTER CONTROL SYSTEM

This section describes the main tasks that computer control systems perform. Figure 1 is a broad illustration of a computer control system.

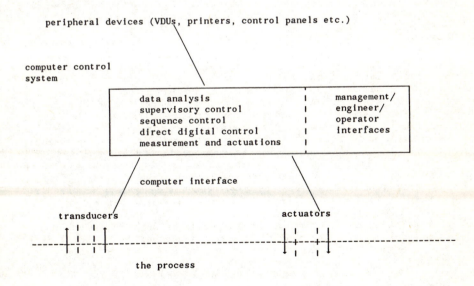

Fig.1   Schematic diagram of a computer control system

Signals are monitored from and are supplied to the controlled process. Collected data are analysed at various levels to provide plant adjustments for automatic control or to provide information for managers, engineers and operators.

2.1 Measurement

The plant operation is monitored using transducers. These are devices that generate an electrical signal that is proportional to a physical quantity on the plant that is to be measured (ie. temperature, pressure, flow, concentration, etc). The transducer signals are connected to the measurement interface of a computer system, which is usually standardised to accept signals of one particular kind. A widely adopted standard is that the 0 to 100% range of any particular measurement converts to 4 to 20mA or 1 to 5V. For example if a particular transducer monitors flow over the range 10 litres/hr to 50 litres/hr, then the transducer would be calibrated so that 10 litres/hr suplies 4mA and 50 litres/hr supplies 20mA to computer interface. A different standard may sometimes be adopted for temperature measurements taken via thermocouple or resistance thermometer transducers.

Another variety of measurement concerns the status of various aspects of the controlled process. Is a valve open or closed? Is a vessel full of liquid? Is a pump switched on? Such information would be supplied to the computer in digital form, perhaps by the opening and closing of relay switches or by the level of a TTL voltage.

The computer may also monitor digital data directly, via a serial or parallel data communications link. Many transducers now utilise microprocessors, for example liquid concentration analysers. Typically a microprocessor might apply statistical analysis to extract the required information from the monitored plant signal. Given a numerical result it is then straightforward to effect the transfer to the control computer.

The control computer maintains a record of all of the measurements. Periodically this record is updated by scanning each of the signals connected to the interface. Each measurement may be referenced, say, twice each second. At these instances, the electrical signal is converted to a numerical equivalent by an analogue-to-digital converter (ADC). If the ADC converts to a 10-bit binary number, which is commonly the case, then 4mA returns 0 and 20mA returns 1023 for the standard signal. To be meaningful, the returned number must be scaled to engineering units. For a transducer with a range of 10 to 50 litres/hr, 0 scales to 10 and 1023 to 50. An intermediate value y ($0 < y < 1023$) scales to $(y/1023) \times 40 + 10$. The incoming signal sometimes has to be linearised; certain flow transducers require the square root of a signal to be evaluated after measurement; thermocouples generate a voltage that relates to temperature via a polynomial formula which has to be evaluated after measurement.

If a measurement signal has a lot of noise superimposed upon it, digital filtering may be applied to smooth the signal. In such a case, if a representative value is required every second, then the signal must be sampled and processed by the filter more frequently (say every 0.1 secs). The most commonly used filtering procedure employs the first order exponential algorithm.

$$yf_{k+1} = a.yf_k + (1-a)y_{k+1} \quad k = 0, 1, 2 \ldots.$$

where yf_k is the filtered value at instant k, y_{k+1} is the measurement at instant k+1 and a = exp $(-T/\tau)$ with T = the interval k to k+1 (0.1 secs) and τ is the filter time constant. The selection of the times T and τ depends upon the frequency at which the filtered measurement is wanted (1 second in this example) and the frequency and amplitude of the noise on the measurement.

Most measurements taken by the process control computer will also be checked to ascertain whether or not the plant is in a safe state of operation. Two sets of limits usually relate to each measurement. If the inner range is exceeded, a warning status is established. If the outer range is exceeded then a "red alert" condition applies. Automatic checking for such emergency conditions is a very important feature of the on-line computer.

Therefore, for on-line control, each measurement must be converted, possibly filtered, possibly linearised, scaled to engineering units and, finally, checked against alarm limits. The intervals between measurement samples may be quite long in some cases (many seconds) but in other cases many samples may have to be taken each second. If there are a lot of different measurements to be taken, and some installations require many hundreds, there is considerable data processing necessary to bring all of the measurements to a form meaningful for inspection by humans and for use in controllers. Careful choice of sampling intervals, eg, by not monitoring signals more frequently than necessary, can reduce the computational burden. The selection of such intervals also requires care to ensure that problems such as signal aliasing do not arise.

2.2 Actuation

Control of plant is usually achieved by adjusting actuators such as valves, pumps and motors etc. The control computer may generate a series of pulses to drive the actuation device to its desired setting. In such a case, the drive signal would be generated as a relay contact closure or a change in voltage level. Alternatively, a voltage that is proportional to a desired setting may be produced by a digital-to-analogue converter (DAC). An actuation device will often supply a measurement back to the computer so that it is possible to check whether or not an actuation command has been implemented. The computation associated with actuation is usually small, however, some pulse drive actuators require frequent associated measurement to determine when the desired setting has been reached. In this case a more significant computational burden can be incurred.

2.3 Direct Digital Control (DDC)

Conventional analogue electronic control systems employ the standard three-term algorithm

$$u = K_p \left(e + \int \frac{edt}{T_i} + T_d \cdot \frac{de}{dt} \right) \qquad (1)$$

with e = r −y, y is the measurement, r is the reference or set point, e is the error, K_p is the controller gain ($1/K_p$ is the proportional band), T_i is the integral action time and T_d the derivative action time. There are variations on this form. Very few controllers actually use derivative action. If it is used de/dt is sometimes replaced by dy/dt to avoid differentiation of set point.   Nearly all industrial control problems are solved by application of this algorithm or close variations to it.

Most DDC systems utilise a difference equation equivalent to the above algorithm. If the computer recalculates the actuation signal u every T seconds, then the most simple numerical approximation employs

$$\frac{de}{df} \simeq \frac{e_k - e_{k-1}}{T} \text{ and } \int e.dt = \Sigma e_k.T \qquad k = 0, 1, 2, \ldots\ldots \qquad (2)$$

with the interval k to k+1 equal to T seconds.
The control equation then results as

$$u_k = K_p \left\{ e_k + \frac{T.s_k}{T_i} + \frac{T_d}{T} (e_k - e_{k-1}) \right\} \qquad (3)$$

with $s_k = s_{k-1} + e_k$ being the sum of errors.
The advent of cheap microprocessors has made it very simple to programme and implement this control equation. However, there are many traps into which the self-taught control engineer can fall. One of the most significant is associated with saturation of the actuation signal. This signal must be considered to lie within a defined range Umin/Umax. If the control signal saturates at either extreme careful consideration has to be given to the integral sum s_k at this stage. If the summation procedure were to continue unchecked s could take up a large and unrepresentative value which could lead to much degraded control system performance thereafter. Special procedures for integral desaturation have been developed to accompany the controller. These ensure that the actuation signal emerges from saturation at a timely moment so that an effective control system response is generated.

DDC may be implemented on a single loop basis by a single microprocessor controller or by a larger computer which could implement upwards of a hundred control loops. The total control system for an industrial process can become quite complex. DDC loops might be cascade connected, with actuation signals of particular loops acting as set points for other loops. Signals might be added together (ratio loops) and conditional switches might alter signal connections. Figure 2 is a diagram of a control system that is widely used for the management of boiler systems. The pressure of steam produced by a boiler is to be controlled.   This is achieved by regulating the supply of fuel oil to the burner. However, a particular mix of fuel and air is required to ensure efficient and non-polluting combustion. The purpose of this illustration is not to discuss how the control system achieves its objective, but rather to indicate some of the elements that are required for industrial process control. Such elements must be available as features of a process control computer.

Referring to Figure 2 the steam pressure control system generates an actuation signal that is fed to an automatic/manual bias station. If the latter is selected to be in automatic mode, the actuation signal is transmitted through the device; if it is in the manual bias mode, the signal that is transmitted is one that has been manually defined (eg, by typing in a value on a computer keyboard). The signal from the bias station is connected to two units, a high signal selector and a low signal selector. Each selector has two input signals and one output. The high selector transmits the higher of its two input signals and the low selector the lower of its two input signals. The signal from the low selector cascades a set point to the DDC loop that controls the flow of oil. The signal from the high selector cascades a set point to the DDC loop that controls the flow of air. Finally, a ratio unit is installed in the air flow measurement line. A signal that is generated from another controller is added to the air flow signal prior to its being fed to the air flow controller. This other controller is one that monitors the combustion flames directly, using an optical pyrometer for measurement, and thereby obtains a direct measure of combustion efficiency.

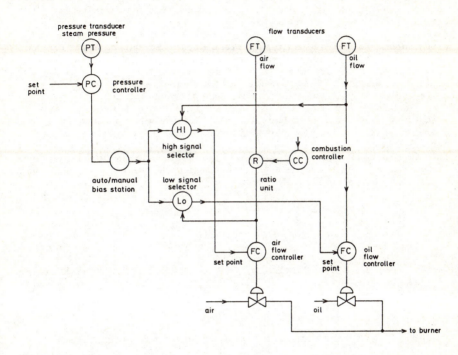

Fig. 2 Boiler presure control system

Another controller that is of value but is not a feature of Fig. 2 is one based on the lead/lag transfer function. This can be used to provide feed-forward compensation so that the effect of disturbances upon the plant, for example because of an alteration in the environmental conditions, is minimised.

Hence there is much more to computer control than simple DDC. The boiler examples are probably more complex than many industrial schemes although the use of additional signal processing over and above simple DDC is very common.

DDC is not restricted to the three-term algorithm described above, although the latter is almost universally used in the process industries. Algorithms based upon, for example, z-transform design techniques can be equally effective and a lot more flexible than the three-term controller. However, the art of tuning three-term controllers is so well established amongst the control engineering fraternity that new techniques are slow to gain ground. The fact that three-term control copes perfectly adequately with 90% of all control problems is, also a deterrent to general acceptance of new concepts for control system design.

In the last 2 to 3 years most major manufacturers have developed an auto-tuning feature for their microprocessor/computer controllers. These have the capability to automatically calculate the parameters Kp, Ti and Td for suitably controlled performance. Such self-tuning control systems are in the early days of industrial use and, although currently of only limited market penetration, application is likely to become commonplace before too long. The most common approach for tuning is to characterise the plant dynamics by some simple first or second order transfer function. Usually the process is subjected to some small disturbances to provide suitable plant input and output data for such characterisation. Simple analytical techniques supported by a few "expert system" rules seem to provide the most robust approach for automatic design. These, in effect, implement similar procedures to those employed by an engineer when a loop is manually tuned.

A problem with the new tuners is that they are often used in situations where engineers have found it difficult to adequately tune a loop by manual techniques. Such situations are frequently inappropriate for the use of three term control, perhaps because of the presence of large time delays or because of large disturbances impacting from elsewhere. Thus the tuners need to be used with caution and should not be seen as the panacea to all control engineering difficulties.

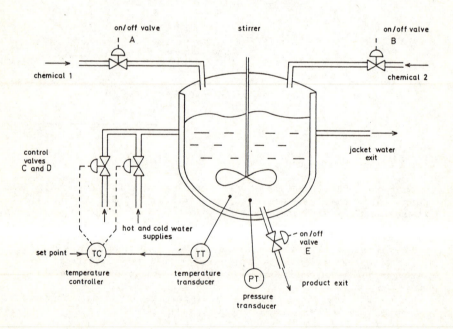

Fig.3 A chemical reactor

2.4 Sequence Control

Many industrial processes are required to be automatically sequenced through a number of stages during their manufacturing operations. For example, consider the manufacture of a chemical that is produced by reacting together two other chemicals at a particular temperature. Fig. 3 illustrates a typical plant arrangement for this purpose. The chemicals react in a sealed vessel (the reactor). The contents are temperature controlled by feeding hot or cold water to the water jacket that surrounds the vessel. This water flow is manipulated by adjusting the control valves C and D. On/off valves A, B and E in the chemical supply and vessel exit pipelines are used to regulate the flow of material into and out of the vessel. The temperature of the vessel contents and the pressure at the bottom of the vessel are monitored.

The manufacturing procedure for this plant might involve the following stages of operation.

1 Open valve A to charge chemical 1 to the reacting vessel.
2 Check the level of chemical in the vessel (by monitoring the pressure vessel). When the required amount of chemical has been charged, close valve A.
3 Start the stirrer to mix the chemicals together.
4 Repeat stages 1 and 2, with valve B, in order to charge the second chemical to the reactor.
5 Switch on the three-term controller, and supply a set point so that the chemical mix is heated up to the required reaction temperature.
6 Monitor the reaction temperature. When it has reached set point, start a timer to time the duration of a reaction.
7 When the timer indicates that the reaction is complete, switch off the controller and open valve C to cool down the reactor contents. Switch off the stirrer.
8 Monitor the temperature. When the contents have cooled, open valve E to remove the product from the reactor.

When implemented by a computer, all of the above decision-making and timing is based upon software. For large chemical plants, such sequences can become very lengthy and intricate, especially when a plant involves many reaction stages. For the most efficient plant operation, a number of sequences might be in use simultaneously (eg in the context of the above example a number of reactions might be controlled at the same time). Very large process control computers are often dedicated almost exclusively to supervising such complex sequence-control procedures.

2.5 Supervisory Control Systems

Supervisory control systems are used to specify or optimise the operation of the set of DDC (or conventional analogue control) systems that are controlling a plant. For example, the objective of a supervisory system might be to minimise the energy consumption of a plant or to maximise its production efficiency. A supervisory system might compute the set points against which the plant control systems are to operate or it might re-organise the control systems in some way.

A simple example of where a supervisory control scheme can be utilised is illustrated in Fig. 4. Two evaporators are connected to operate in parallel. Material in solution is fed to each evaporator.

The purpose of the plant is to evaporate as much water from the solution as possible. Steam is supplied to a heat exchanger linked to the first evaporator. Steam for the second evaporator is supplied from vapours boiled off at the first evaporation stage. To achieve maximum evaporation the pressures in the evaporation chambers must be as high as safety considerations will allow. However, it is necessary to achieve a balance between the two evaporators. For example, if the first evaporator is driven flat out then this might generate so much steam that the safety thresholds for the second evaporator are exceeded. A supervisory control scheme for this example will have the task of balancing the performance of both evaporators so that, overall, the most effective rate of evaporation is achieved.

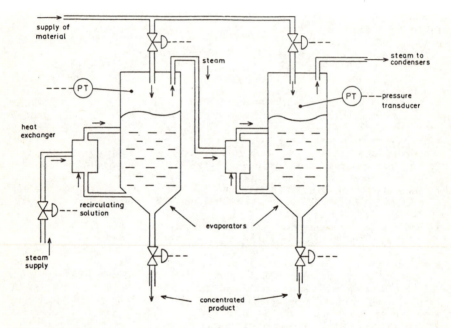

Fig.4   An evaporation plant

In most industrial applications supervisory control, if used at all, is very simple and is based upon knowledge of the steady-state characteristics of the plant to define the required plant operating status. In a few situations, very sophisticated supervisory control algorithms have proved beneficial to plant profitability. Optimisation techniques using linear programming, gradient search methods (hill climbing), advanced statistics and simulation have been applied. In association with these techniques, complex non-linear models of plant dynamics and economics have been solved continuously in real time in parallel with plant operations, in order to determine and set up the most effective plant operating point. An example is the processing of crude oil by distillation. The most profitable balance of hydrocarbons can be produced under the direction of a complex supervisory control system.

Quite recently, expert system approaches for the solution of supervisory control system problems have begun to be given serious consideration. The expert system methodology is to empirically build up a set of logical rules that reflect the decision making that would be invoked by the skilled human operator. The cement industry in particular has given particular consideration to expert systems for the supervision of the operation of cement kilns. There are a number of reported successful implementations. The approach may employ the use of fuzzy logic to add a low degree of artificial intelligence to the decision making process. This is probably an exciting area for future development and application.

Other recent advances have taken place in the Petrochemicals Industry. A technique termed QDMC (Quadratic Dynamic Matrix Control) has found successful application in solving the "Dynamic Constraint Optimisation Problem". The problem is to transfer to a process from one state to another, subject to an economic cost requirement, without violating any process constraints en route. It employs Quadratic Programming in conjunction with linear prediction theory.

2.6 The engineer and the process control computer

The control engineer has the task of specifying the various roles of the process-control computer and of implementing the specification. His duties may be itemised as follows (presuming the decision has been made as to the most suitable computer system for the job in hand).

1   To define measurements and actuations and set up scaling and filter constants, alarm and actuation limits, sampling intervals etc.
2   To define the DDC controllers, the interlinking/cascade connections between them and any other elements within the control system configuration.
3   To tune the above control systems, ie select appropriate gains, so that they perform according to some desired specification.
4   To define and programme the sequence control procedures necessary for the automation of plant operation.
5   To determine and implement satisfactory supervisory control schemes.

The control engineer may also have the job of determining how the plant operator is to use the computer system in the day-to-day running of the plant.

Clearly, for a large application, all of these duties would be beyond the scope of a single person and a team of people would be involved, one of whom would most certainly be a computer programmer. The latter phase, implementation of supervisory control, could be a project of many months if some of the more sophisticated techniques above mentioned are utilised.

The programming effort involved in establishing a complete working computer-control system could be considerable if the engineer had to start from scratch. However, process control applications have many aspects in common and standard packages of computer software are now available with many computing systems, thus minimising the effort required to establish a working control system. Facilities are available to permit the engineer to translate directly from specification charts and control system diagrams, such as that of Fig.2, to a computer based implementation. The engineer defines the database but does not have to drite the software for the DDC systems. The standard software is usually sold with the computer system to form a complete process-control package.

Software packages for sequence control and supervisory control must be a lot more flexible than those for DDC. Sequencing and supervisory requirements will differ greatly from plant to plant whereas DDC configurations utilise a very limited range of standard operations. The standard packages therefore provide the engineer with a higher level language to programme the required seqence of commands. A variety of such languages exist. The most common are very similar to BASIC, with additional features to facilitate the real-time aspects and communications with the plant interface. Ladder Logic is also very common, particularly relating to PLC systems (Programmable Logic Control). If the language is interpretive (ie lines of programme are compiled at the time of execution), it is often possible to build up a sequence procedure while the computer is online to the plant. Supervisory control might require extensive calculations for which interpretive operation could be too slow and cumbersome. In this case, a language such as Fortran might be utilised and compilation would be necessarily offline. For the simpler supervisory schemes, the BASIC type languages are perfectly adequate.

2.7 Facilities for plant operator and plant manager

The plant operator must be provided with facilities that permit the straightforward operation of the plant on a day-to-day basis. The operator requires to be presented with all information relevant to the current state of operation of the controlled process and its control systems. In addition it is necessary for him to be able to interact with the plant, for example to change set points, to manually adjust actuators, to acknowledge alarm conditions etc.

A specially designed operator's control panel is a feature of nearly all computer control systems. Such a panel would typically consist of special keyboards, perhaps tailor-made for the particular plant that is controlled, and a number of display screens and printers. The video displays permit the operator to inspect, at various levels of detail, all monitored areas of the plant. The standard software packages supplied with the computer control systems normally provide a range of display formats that can be used for the representation of information. Typically, these might be an alarm overview display presenting information relating to the alarm status of large groups of measurements; a number of area displays presenting summaries of details concerning the control systems associated with particular areas of the process; and a large number of loop displays, each giving comprehensive details relating to a particular control loop. The control engineer selects the parameters that are to be associated with the individual displays, as part of the procedure of defining the data base for the computer system. The display presentations might be in the form of ordinary print-outs, or trend graphs for measurements, or schematic diagrams (mimics) of plant areas, with numerical data superimposed at appropriate locations. Most installations will utilise colour-graphic displays.

The above standard displays will not suit all requirements. For example, sequence control procedures might require special display presentation formats so that the operator can establish and interact with the current stage of process operation. Such displays would most likely be produced by programming them specially using the BASIC type language referred to above.

The special operator keyboards are usually built to match the standard display structures (for example by specifying particular keys to be associated with the selection of displays for particular plant (areas). It is thereby straightforward for the operator to quickly centre upon aspects of interest. Commonly, if a particular control loop is pinpointed on the display, perhaps by use of a cursor or a light pen, then this will permit the operator to make direct adjustments to that loop using special purpose keys on a keypad.

The plant manager requires to access different information from the process control computer than the operator. He will need hard copy print-outs that provide day-to-day measures of plant performance and a permanent record of the plant operating history. Statistical analysis might be applied to the plant data prior to presentation to the manager so that the information is more concise and decisions are therefore more straightforward to make. The manager will be interested in assessing performance against economic targets, given the technical limitations of the plant operation. He will determine where improvements in plant operation might be possible. He will be concerned, along with the control engineer, with the operation of the supervisory control systems and will set the objectives for these top-level control systems.

3 HARDWARE FOR THE COMPUTER CONTROL SYSTEMS

In this chapter it has been assumed so far that the process control computer is a single hardware unit, that is, one computer performing all of the tasks itemised in section 2. This was nearly always the case with early computer control systems. The recent rapid developments of solid state and digital technology have led to a very different approach to the hardware configuration of computer control systems.

Fig. 5 illustrates a typical arrangement that might now apply with many modern systems. The tasks of measurement, DDC, operator communications and sequence control etc., are distributed amongst a number of separate processing units, each of which will incorporate a microcomputer of one form or another. The microcomputers are linked together via a common serial communications highway and are configured in a hierarchical command structure.

Fig. 5 indicates 5 broad levels of division in the structure of the hierarchy of microcomputer units. These correspond in the main with functions described in section 2.

Level 1:- all components and plant interfacing associated with measurement and actuation. This level provides the measurement and actuation data base for the whole system.
Level 2:- all DDC calculations.
Level 3:- all sequence control calculations.
Level 4:- operator communictions.
Level 5:- supervisory control.
Level 6:- communications with other computer systems.

The boundaries between the levels do not have to be rigid, for example a unit for DDC might also perform some sequence control or might interface directly to the plant. PLC systems, for example programmed via ladder logic, generally encompass levels 1, 2 and 3 and connect via data link to operator and supervisory systems.

The units for operator communications will drive the operator control
panels and the associated video displays. The levels within the system
define a command structure. Thus the microcomputer for supervisory
control may direct the sequence control computers which, in turn,
provide set points for the DDC computers etc.

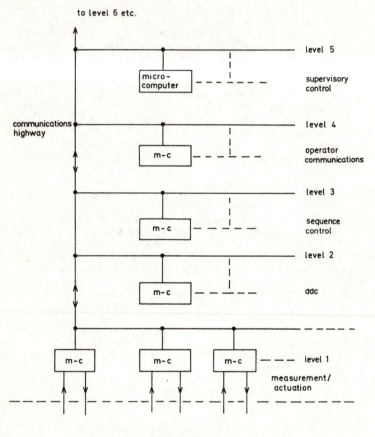

Fig.5 A distributed and hierarchical microcomputer system

The major features and advantages of this distributed/hierarchical
approach are:

i) If the computational tasks are shared betwen processors, then system
capability is greatly enhanced. The burden of computation for a single
processor becomes very great if all of the described control features
are included. For example, one of the main computing loads is that of
measurement scanning, filtering and scaling. This is not because any

one calculation is onerous but rather because of the large number of signals involved and the frequency at which calculations have to be repeated. Separation of this aspect from DDC, if only into two processors, greatly enhances the number of control loops that can be handled. The DDC computer will collect measurements already processed, via the communications link, at a much lower frequency than that at which the measurement computer operates.

ii) The system is much more flexible than a single processor. If more control loops are required or an extra operator station is needed, then all that is necessary is to add more boxes to the communications link. Of course, the units already in the system must be updated to "be aware" of the additional items.

iii) If any unit should fail, the implications are not catastrophic, only one portion of the overall system will be out of commission, not the whole assembly.

iv) It is much easier to make software changes to the distributed system. For example, if a supervisory control programme is to be altered, then only the associated microcomputer need be called off-line. The risks of causing total system failure because of computer programming faults are very much reduced.

v) The units in the system can become standardised. This leads to a much lower cost facility overall. Thus, typically, a microprocessor for DDC might be standardised to cater for 16 loops. All of the necessary aspects such as gains, limits, set points, etc., would be communicated from elsewhere via the highway. An application requiring 80 control loops would therefore utilise 5 such DDC boxes. In fact, at the extreme low end of the computer control market there are microprocessor units that implement a single control loop. These, to all external appearances, apart from their data link facility, look very similar to conventional electronic three-term controllers.

vi) The interlinking of the microcomputer units by a serial highway means that they can be dispersed over quite a wide area. The highway may easily stretch for a mile or even more of telecommunications devices are used. An advantage is that it makes it unnecessary to bring many cables carrying transducer signals to the control rooms. The measurement microprocessors can be sited close to the source of the signals (ie near the plant) and then only a serial link need be taken back to the control room. Another advantage is that it is straightforward to set up multiple operator control terminals at different sites in the factory. Hierarchical distribution of process-control computers is not an innovation with microcomputers. Very large chemical processes, or even complete factories, for which a large capital outlay for control systems can be justified, have employed the above principles. It is the reductions in cost and size of the computing facilities that have made the technology the most attractive for the majority of industrial control applications today.

4   INTEGRITY OF COMPUTER CONTROL SYSTEMS

One of the main barriers to the use of computer control systems, apart from expense, has been mistrust of the computer. A conventional instrumentation and control system uses many individual units and the failure of one or a few can be tolerated without having to shut down plant operations. If all of these units are replaced by a single computer then a break down of the computer can result in a complete loss of all control systems, with unfortunate consequences. The typical mean-time-between-failures for early computer control systems has been in the region of 3 to 6 months. Many applications have therefore only used a computer for aspects such as sequence and supervisory control for which certain degrees of failure can be tolerated. Continuous feedback control has, in these cases continued to be implemented by electrical or pneumatic instrumentation.

The solution to the problem of computer failure has been to provide back-up systems to take over if a computer failure occurs. The backup systems might be a bare minimum of analogue controllers that are switched in automatically. The difficulty with this approach is that if the computer does not fail for a long time the plant staff might forget how to operate the back-up system. Wise users occasionally switch off the computer deliberately. An alternative back-up mechanism is to duplicate the computer system so that if one fails, another takes over. Certain applications, eg in the nuclear industry, triplicate the control computers. The above options are expensive and it has proved difficult to establish change-over mechanisms that are guaranteed not to disrupt the plant in any circumstances.

The new microcomputer technology has alleviated many of the problems of integrity. Units, such as operator stations, may be duplicated on the data highway at low cost. Units are now programmed to have a self-diagnosis capability so that they can automatically establish and report if faults occur. The most vulnerable aspect of the new aproach is the communication link, if this is broken then all means of access to units on the wrong side of the break are lost, and for this reason, the communications link is often duplicated. Change over between links is automatic if a fault is detected. Some manufacturer's systems permit duplication to be extended to cover almost every unit within the distributed network. Mean-time-between-failures that would result in a total loss of plant control, are quoted at between 50 and 100 years. Needless to say, at such levels of duplication the systems cease to be cheap.

5   THE ECONOMICS OF COMPUTER CONTROL SYSTEMS

Before the advent of microprocessors, computer control systems were very expensive. When a new industrial process was being designed and built, or an old one being re-instrumented, a strong case had to be presented to use a computer in favour of conventional instrumentation.

In some cases computers have been used because, otherwise, plants could not have been made to work profitably. This is particularly the case with large industrial processes that require the application of complex sequencing procedures. The computer system permits repeatability in quality that is essential, for example, with manufacturing plants in the pharmaceutical industry. Flexibility of the computer is also important in these circumstances. Conventional instrumentation is difficult to modify if, for example, a sequencing procedure is to be altered to suit the manufacture of a different product. Reprogramming using a sequence language is comparatively straightforward.

Many large continuous processes (oil refineries, evaporators etc.) are also computer controlled. These processes stay in a steady state of operation for long periods and require little sequencing. They do incorporate many control loops. The usual justification for computer control in these circumstances has been that it will make plant operation more profitable. Such statements are often based upon the argument that even a small increase in production (say 1 to 2%) will more than pay for the computer systems. In the event, it has often been difficult to demonstrate that such improvement has resulted. The author is aware of one major installation where production has declined following the introduction of computers. The pro-computer lobby argue that it would have declined even more if the computer had not been there! The major benefits with continuous processes should arise through a better understanding of the process that invariably follows the application of computer schemes that maintain the plant closer to desired thresholds of operation.

The scene relating to the justification for computer control systems has now changed dramatically. The fall in cost and the improvement in reliability mean that for many industrial applications, both major and minor, it is automatic to install computer controls. In any event, microprocessor units are now cheaper than many of the equivalent analogue instruments, which are now almost obsolute.

6 FUTURE TRENDS

The advent of the microcomputer has probably had more impact upon the discipline of control engineering than any other. Applications are now blossoming in all areas of industry on plants both large and small. The hardware revolution is still taking place but future changes are not likely to be as dramatic as in recent years. The question for the future is: "Will there be an accompanying revolution in the control techniques that are implemented by the new hardware?"

An immense amount of effort has been devoted to academic control engineering research over the past two decades. Numerous novel control procedures have been postulated and proved effective, but only in numerical simulations. So many of the developed procedures could be summarised by the statement "and here is yet another example of a controller for the linear plant described by the state of space equation $\underline{x} = \underline{A} \, x + \underline{B} \, \underline{u}$." There has been very little application of these novel techniques, particularly in the process industries. They have remained very much the domain of the expert in applied mathematics. In industry the three-term controller has continued to be the fundamental control unit and will remain so for a long time to come.

The emphasis in control systems research and development must, and will, change. Much greater emphasis will be given to applying the new techniques and making them work. Computers will be used not only for the implementation of controllers but also to assist with defining controller configurations and with tuning, so that the best controlled performance is achieved. Techniques such as computer-aided design, identification of plant dynamics and adaptive/self-tuning control will become features of standard software packages supplied by the manufacturers of the hardware. Already, as previously mentioned, there is the progress with the provision of automatic tuning aids for conventional control systems. Extensions will progress to cater for control systems that can compensate for time delays, interactions and disturbances. It is possible to conceive of a new kind of box appearing on the data highway of Fig. 5, one for the design of control systems, a facility to assist the control engineer.

The increased emphasis on applications in control engineering research will motivate more problem specific solutions. The preoccupation of recent years with general theory will fade (as it must do anyway since linear control theory must be almost played out by now and extension to non-linear situations almost always requires a special case consideration). Greater attention will be given to consideration of supervisory control methods. This is the area where real financial returns from control systems can be made, by improving the productivity and efficiency of plant operations. The expert system approach would appear to have real potential in this area and is an exciting topic for applications research.

7 CONCLUSIONS

This chapter provides an introduction to the current state of the art of computer control systems. Many manufacturers provide standard system packages that implement the described features. Well known names such as Honeywell, Taylor, and Fisher cater for the distributed and hierarchical approach described in section 3. (TDC 3000, MODIII and PROVOX respectively). Firms such as Turnbull Control Systems and Bristol Automation are marketing attractive microprocessor based systems that are beginning to take on the same capabilities. Companies such as Allen Bradley, AEG (MODICON), and Square D market PLC systems which are now commonplace in industry worldwide. Even the microprocessor companies themselves, such as Texas Instruments are now producing their own variations of distributed process control systems. The market is lively and competitive and costs are falling in line with the general downward trend of the price of computing equipment. Computer control is becoming commonplace.

Applications of new concepts in control theory have lagged far behind the boom in hardware. It is anticipated that this will change and that the technology of process control will be refreshed, in its turn, in the next few years. Facilities will become available to assist the control engineer to obtain better performance from the controlled plant than is currently possible unless a great deal of design effort is applied. The use of more sophisticated procedures for improved process control will become more common.

8 BIBLIOGRAPHIC NOTES

1 A best appreciation of the current state of computer systems is obtained by reference to manufacturers' publicity literature. Honeywell have produced a number of pamphlets under the title "An Evolutionary Look at Process Control" in support of their TDC 3000 system. These provide a well presented overview of many of the aspects discussed in this paper. Taylor, Foxboro and Fisher produce similar documentation relating to their new process control systems but of rather less value educationally. Interesting contrasts between the various systems can be established by perusing this literature. A good sample of the capabilities of small systems can be established from the literature of Turnbull Control Systems in support of their series 6000 instruments.

2   For DDC:

WILLIAMS, T.J.: "Direct digital control and its implications for
chemical process control", Derchema Monographien, 1965-6,v53,
No.912-924, pp. 9-43

DAVIES, W.D.T.: "Control algorithms for DDC", Instrumentation
Practice for Process Control and Automation, 1967, v.21, pp.70-77

3   For supervisory control concepts:

LEE, GAINS, ADAMS.: "Computer process control: Modelling and
Optimisation", (Wiley, 1968)

4   For discrete control systems

FRANKLIN, G.F., POWELL, J.D.: "Digital Control of Dynamic Systems"
(Addison Welsey, Reading, Mass., 1980).
CADZOW, J.A., MARTINS, H.R.: "Discrete time and computer control
systems" (Prentice Hall, Englewood Cliffs, 1970).

Index